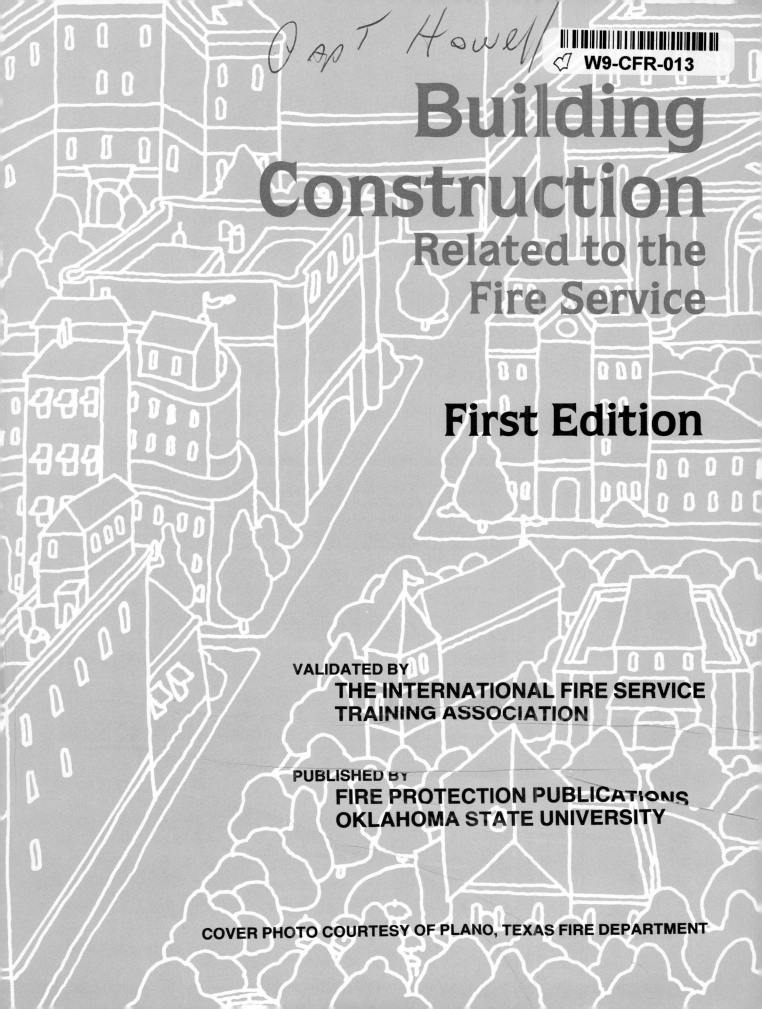

W9-CFR-013

Building Construction
Related to the Fire Service

First Edition

VALIDATED BY
**THE INTERNATIONAL FIRE SERVICE
TRAINING ASSOCIATION**

PUBLISHED BY
**FIRE PROTECTION PUBLICATIONS
OKLAHOMA STATE UNIVERSITY**

COVER PHOTO COURTESY OF PLANO, TEXAS FIRE DEPARTMENT

Dedication

This manual is dedicated to the members of that unselfish organization

of men and women who hold devotion to duty

above personal risk, who count on sincerity of service above

personal comfort and convenience, who strive unceasingly to find

better ways of protecting the lives, homes and property

of their fellow citizens from the ravages of fire and other

disasters ... **The Firefighters of All Nations.**

Dear Firefighter:

The International Fire Service Training Association (IFSTA) is an organization that exists for the purpose of serving firefighters' training needs. Fire Protection Publications is the publisher of IFSTA materials. Fire Protection Publications staff members participate in the National Fire Protection Association and the International Association of Fire Chiefs.

If you need additional information concerning our organization or assistance with manual orders, contact:

Customer Services
Fire Protection Publications
Oklahoma State University
Stillwater, OK 74078-0118
1 (800) 654-4055

For assistance with training materials, recommended material for inclusion in a manual, or questions on manual content, contact:

Technical Services
Fire Protection Publications
Oklahoma State University
Stillwater, OK 74078-0118
(405) 744-5723

First Printing, February 1986
Second Printing, December 1988
Third Printing, November 1993
Fourth Printing, February 1995

Table of Contents

List of Tables

THE INTERNATIONAL FIRE SERVICE TRAINING ASSOCIATION

The International Fire Service Training Association is an educational alliance organized to develop training material for the fire service. The annual meeting of its membership consists of a workshop conference which has several objectives —

... to develop training material for publication
... to validate training material for publication
... to check proposed rough drafts for errors
... to add new techniques and developments
... to delete obsolete and outmoded methods
... to upgrade the fire service through training

This training association was formed in November 1934, when the Western Actuarial Bureau sponsored a conference in Kansas City, Missouri, to determine how all agencies that were interested in publishing fire service training material could coordinate their efforts. Four states were represented at this conference and it was decided that, since the representatives from Oklahoma had done some pioneering in fire training manual development, other interested states should join forces with them. This merger made it possible to develop nationally recognized training material which was broader in scope than material published by an individual state agency. This merger further made possible a reduction in publication costs, since it enabled each state to benefit from the economy of relatively large printing orders. These savings would not be possible if each individual state developed and published its own training material.

From the original four states, the adoption list has grown to forty-four American States; six Canadian Provinces; the British Territory of Bermuda; the Australian State of Queensland; the International Civil Aviation Organization Training Centre in Beirut, Lebanon; the Department of National Defence of Canada; the Department of the Army of the United States; the Department of the Navy of the United States; the United States Air Force; the United States Bureau of Indian Affairs; The United States General Services Administration; and the National Aeronautics and Space Administration (NASA). Representatives from the various adopting agencies serve as a voluntary group of individuals who govern policies, recommend procedures, and validate material before it is published. Most of the representatives are members of other international fire protection organizations and this meeting brings together individuals from several related and allied fields, such as:

... key fire department executives and drillmasters,
... educators from colleges and universities,
... representatives from governmental agencies,
... delegates of firefighter associations and organizations, and
... engineers from the fire insurance industry.

This unique feature provides a close relationship between the International Fire Service Training Association and other fire protection agencies, which helps to correlate the efforts of all concerned.

The publications of the International Fire Service Training Association are compatible with the National Fire Protection Association's Standard 1001, "Fire Fighter Professional Qualifications (1981)," and the International Association of Fire Fighters/International Association of Fire Chiefs "National Apprenticeship and Training Standards for the Fire Fighter." The standards are an effort to attain professional status through progressive training. The NFPA and IAFF/IAFC Standards were prepared in cooperation with the Joint Council of National Fire Service Organizations of which IFSTA is a member.

The International Fire Service Training Association meets each July at Oklahoma State University, Stillwater, Oklahoma. Fire Protection Publications at Oklahoma State University publishes all IFSTA training manuals and texts. This department is responsible to the executive board of the association. While most of the IFSTA training manuals can be used for self-instruction, they are best suited to group work under a qualified instructor.

Preface

Acknowledgement and grateful thanks are offered to the continuing committee members who provided input and worked so diligently on developing this manual.

Chairman
Curtis Holter
Fire Instructors Association of Minnesota
Minnetonka, Minnesota

Vice-Chairman
Max McRae
District Chief
Houston Fire Department
Houston, Texas

The editorial staff also extends its appreciation to the other dedicated committee members who devoted their time and talents toward this manual.

Tom Admire
Jim Badgett
Glenn Boughton
Dave Briley
Bob Caron
James Hebert
Ron Hopkins
Bill Hulsey

Julian King
J. Fred Myers
Wayne Nelson
J.D. Rhyne
Roy Richardson
Lloyd Scholer
Jim Simmons

Gratitude is also extended to the following individuals whose contributions made the final publication of this manual possible: Edward Prendergast, technical editing and photographs; Lynne Murnane and Carol Smith, editorial staff; Scott Stookey, research and writing; Don Davis, Cindy Brakhage, Ann Moffat, Michael McDonald, Desa Porter, and Karen Murphy for their assistance in layout, proofreading, artwork, and phototypesetting of the manual.

Gene P. Carlson
Editor

Glossary

Aggregate — The gravel, stone, sand, or other inert materials used in concrete.

Air-Entrained Concrete — Concrete with air entrapped in its structure to improve its resistance to freezing.

Aluminize — To coat with aluminum.

Anchor — A metal device used to hold down the ends of trusses or heavy timber members at the walls.

ASTM — American Society for Testing and Materials.

A

Backfill — Coarse dirt or other material used to build up the ground level around the foundation walls to provide a slope for drainage away from the foundation.

Backsplash — The vertical surface at the back of a countertop.

Baluster — The vertical member supporting a handrail.

Balustrade — The entire assembly of a handrail and its supporting members (newel posts and balusters).

Batt Insulation — Blanket insulation cut in widths to fit between studs and in short lengths to facilitate handling.

Bay — The space between bent, beams, or between rows of columns considered in transverse planes.

Beam — A structural member subjected to loads perpendicular to its length.

Bench Mark — A mark on some object firmly fixed in the ground from which distances and elevations are measured.

Bent — Supporting legs of a bridge in a plane perpendicular to its length.

Bifold Doors — Doors designed to fold in half vertically.

Blanket — A thick layer of insulating material between two layers of heavy waterproof paper.

Bolster — A device made of bent wire used to hold several reinforcing bars in position.

B

Brick Veneer — A single thickness of brick wall facing placed over frame construction or masonry other than brick.

Bridging — Diagonal bracing between joists.

Buttress — A structure projecting from a wall designed to receive lateral pressure action at a particular point.

C

Camber — A low vertical curve placed in a beam or girder to counteract deflection caused by loading.

Cant Strip — An angular board installed at the intersection of a roof deck and a wall to avoid a sharp right angle when the roofing is installed.

Cantilever — Projecting beam or slab supported at one end.

Calk (Caulk) — A nonhardening paste used to fill cracks and crevices.

Carriage (also Stringer) — The main support for the stair treads and risers.

Chair — A device of bent wire used to hold reinforcing bars in position.

Chord — Main members of trusses as distinguished from diagonals.

Cladding — The exterior finish or skin.

Clerestory — A windowed space which rises above lower stories to admit air, light, or both.

Coffer Dam — A watertight enclosure usually made of sheet piling which can be pumped dry to permit construction inside.

Column — A vertical supporting member.

Combplate — A cleated plate located adjacent to the point where escalator steps move below the floor surface.

Common Brick — A fired clay brick with a plain, unfinished surface.

Compression — A force which tends to push the mass of a material together.

Convenience Outlet — An electrical outlet which can be used for lamps and other appliances.

Cope Steel — **Construction:** to cut a flange section in order to avoid interference with other member. Carpentry: to cut a piece of molding to fit another piece joining it at an inside corner.

Cornice — A horizontal projection which crowns or finishes the eaves of a building.

Curing — Maintaining conditions to achieve proper strength during the hardening of concrete.

D

Datum Point — A point of reference established by a city from which levels and distances are measured.

Dead Load — The load on a structure due to its own weight and other fixed weights.

Deformations — Projections on the surface of reinforcing bars to prevent the bars from slipping through the concrete.

Draft Curtain — A noncombustible barrier extending down from the ceiling to impede the flow of heat.

Drop Panel — A type of concrete floor construction in which an area above each column is dropped below the bottom level of the rest of the slab.

Drywall — A system of interior wall finish using sheets of gypsum board and taped joints.

E

Efflorescence — Crystals of salt appearing as a white powder on the surface.

Elevation — Measurement: the height of point above sea level or some datum point. Drafting: the drawing or orthographic view of any of the vertical sides of a structure or vertical views of interior walls.

Expansion Joint — A flexible joint in concrete used to prevent cracking or breaking because of expansion and contraction due to temperature changes.

Extrude — To shape heated plastics or metal by forcing them through dies.

F

Fan Light — A semicircular window, usually over a doorway, with muntins radiating like the ribs of a fan.

Fascia — A flat vertical board located at the outer face of a cornice.

Fillet Weld — A weld made in the interior angle of two pieces placed at right angles to each other.

Fire Cut — An angled cut made at the end of a joist or wood beam which is inserted into a masonry wall.

Flashing — Sheet metal used in roof and wall construction to keep water out.

Flat Slab — A type of concrete floor construction which provides a flat surface for the underside of the floor.

Flight — A series of steps between two landings.

Foil Back — Blanket or batt insulation with one surface faced with metal foil which serves as a vapor barrier and heat reflector.

Footing — That part of the building which rests on the bearing soil and is wider than the foundation wall. Also the base for a column.

Front Stringer — The stringer that supports the side of the stairs with a balustrade.

Furring — Wood strips fastened to a wall, floor, or ceiling for the purpose of attaching covering material.

G

Gage Lines — (Steel construction) Lines parallel to the length of a member on which holes for fasteners are placed. The gage distance is the normal distance between the gage line and the edge or back of the member.

Girder — A large, horizontal, structural member used to support the ends of joists and beams.

Grade Beam — A concrete wall foundation in the form of a strong reinforced beam which rests on footings or caissons spaced at intervals.

Gusset Plate — A plate that is used to connect the members of a wood or metal truss.

H

Handrail — The top piece of a balustrade that is grasped when ascending or descending a stairway. Handrails may be attached to the wall in closed stairways.

Head — The top of a window or door frame.

Hip — The junction of two sloping roof surfaces forming an exterior angle.

Hydration — The chemical process in which concrete changes to a solid state and gains strength.

I

Insert — (Concrete) A metal anchor placed in a concrete wall or beam to which shelf angles are attached.

Insulating Glass — Two panes of glass separated by an air space and sealed around the edge.

J

Joist — A framing member which directly supports the floor.

K

Kiln Dried — A term applied to lumber which has been dried by artificially controlled heat and humidity to a prescribed moisture content.

Kip — One thousand pounds.

Kraft Paper — A strong brown paper made of sulfate pulp.

L

Lagging — Heavy sheathing used in underground work to withstand earth pressure.

Lag Screw — Large wood screw with a hexagonal or square head for turning with a wrench.

Laminated Plastic — Sheet material made of lamination of cloth or other fiber impregnated with plastic and brought to the desired thickness or shape with heat and/or pressure.

Lamination — (Heavy Timber) One of several layers of lumber making up a laminated beam.

Landing — The floor on each story where a flight of stairs begins and ends.

Ledger — (Concrete construction) A horizontal framework member, especially one attached to a beam side that supports the joists.

Lift — (Concrete work) The dimension from the top of one pouring of concrete in a form to the top of the next pouring, e.g., "Pour concrete in 8-inch (203 mm) lifts."

Lift Slab — A system of concrete construction in which the floor slabs are poured in place at the ground level and then lifted to their position by hydraulic jacks working simultaneously at each column.

Light — A pane of glass.

Lintel — A support for masonry over an opening, usually made of steel angles or other rolled shapes singularly or in combination.

Live Load — All furniture, people, or other movable loads not included as a permanent part of the structure.

M

Mill — One thousandth of an inch (.001 inch [.0254 mm]).

Modular Measure — A system of measurement designed to have the parts fit together on a grid of a standard module. The module is four inches (102 mm).

Mullion — The vertical division between multiple windows.

Muntin — The small members dividing the glass lights in a window sash.

N

Neat Cement — A pure cement uncut by a sand mixture.

Newel — The outer posts of balustrades and the stiffening posts at the angle and platform of stairways.

Nosing — The projection of a tread beyond the face of a riser.

O

Open Stringer — A stringer that is notched to follow the lines of the treads and risers.

Orientation — (1) The direction in which a building faces. (2) Relating blueprints to the actual structure with respect to direction.

Overburden — Loose earth covering a building site.

P

Panel Points — Points where the load of roof panels are transferred to trusses.

Parapet — A low wall at the edge of a roof.

Parquet Flooring — Usually of wood, laid in an alternating or inlaid pattern to form various designs. The flooring strips may be glued together to make square units.

Partition — An interior wall which separates a space into rooms.

Penthouse — A room or building built on the roof, usually to cover stairways, house elevator machinery, contain water tanks and/or heating and cooling equipment.

Pier — A supporting section of wall between two openings. Also a short masonry column.

Pilaster — A rectangular masonry column built into a wall.

Pitch — (Steel construction) Spacing between rivet centers. General construction: the slope of a roof expressed as a ratio of rise to span.

Plancier — A board which forms the underside of an eave or cornice.

Plat — A drawing of a parcel of land giving its legal description.

Plate — (Frame construction) The top or bottom horizontal structural member of a frame wall or partition.

Platform — Intermediate landing between floors to change the direction of a stairway or to break up excessively long flights.

Polyethylene Membrane — A type of plastic sheet used for waterproofing.

Porcelainize — To coat with a ceramic material.

Prestressing — A means whereby the reinforcing bars in concrete beams are placed in tension before the concrete is poured so that the member will develop greater strength after the concrete has set.

PSI — Pounds per square inch.

Purline — A horizontal member between trusses which supports the roof.

R

Rabbet — A groove cut in the surface or on the edge of a board to receive another member.

Rafter — A beam that supports a roof.

Reglet — A water seal for roofing in a parapet wall. (Also the masonry units.)

Ribbon — (Frame construction) A narrow strip of board cut to fit into the edge of studding to help support joists.

Rigid Conduit — Nonflexible steel tubing used for the passage of electrical conductors.

Rise — The vertical distance between treads or of the entire stairs.

Riser — General: the vertical part of a stair step. Plumbing: a vertical water supply line.

Rolled Shape — A structural steel member made by passing a hot steel billet between shaped rollers until it reaches the required shape and dimensions.

Romax — A trade name for nonmetallic shielded cable.

Rowlock — A method of laying brick on edge so that the vertical ends appear in the face of the wall.

S

Salamander — A portable heater used on construction jobs.

Sanitary Sewer — Underground pipe used to carry off waste from water closets and from other drains.

Sanitary Tee — A soil pipe fitting with a side outlet to form a tee shape. The side outlet extends with a smooth radius to permit unhampered flow in the fitting.

Sawtooth Roof — A roof with a profile of vertical and sloped surfaces resembling a saw.

Screed — Two or more strips set at desired elevation so that concrete may be leveled by drawing a straightedge over their surface; also the straightedge.

Setback — The distance from the street line to the front of a building.

Shear Stress — The stress resulting when two forces act on a body in opposite directions in parallel adjacent planes.

Sheathing — The covering applied to the framing of a building to which siding is applied.

Shelf Angles — Angles fastened to the face of a building at or near floor levels to support masonry or wall facing materials.

Shoe — A metal plate device used at the bottom of heavy timber columns.

Shoring — Temporary support for formwork.

Sill — Frame construction: the bottom rough structural member which rests on the foundation. General construction: the bottom exterior member of a window or door or the masonry below.

Site — The location of a building or construction.

Slab — Steel construction: a heavy steel plate used under a steel column. Reinforced Concrete construction: the reinforced concrete floor itself.

Sleeve — A tube or pipe extending through a floor slab to provide openings for the passage of plumbing and heating pipes to be installed later.

Soffit — A lower horizontal surface such as the undersurface of eaves or cornice.

Soil Stack — A vertical pipe which runs from the horizontal soil pipe to the house drain to carry waste, including that from water closets.

Spalling — Occurs when excess moisture trapped within the cement of the concrete expands. This expansion of the moisture results in tensile forces within the concrete, causing it to break apart.

Spandrel — That part of a wall between the head of a window and the sill of the window above it.

Strap — (Heavy timber construction) A metal piece used to hold joints in heavy timber construction together.

Strata — A sheetlike mass of rock or earth of one kind found in layers between layers of other kinds of material.

Stressed Skin — The outer surface of a structure when it provides lateral support.

Stringer — Reinforced Concrete construction: horizontal structural member supporting joists and resting on vertical supports. General construction: the member on each side of a stair which supports the treads and risers.

Stud — Vertical structural uprights which make up the walls and partitions in a frame building.

T

Temperature Bar — (Reinforced Concrete construction) Reinforcing bar used to counteract stress caused by temperature changes.

Tensile Stress — A stress in a structural member which tends to stretch the member or pull it apart.

Tier — A horizontal division of a multistory building, usually the stories in a steel frame building.

Tie — Masonry veneer: a metal strip used to tie masonry wall to the wood sheathing. Concrete formwork: device used to tie the two sides of a form together.

Tongue and Groove — A projection on the edge of a board that fits into a recess in an adjacent board.

Tread — The horizontal face of a step.

Tremie — A chute used to deliver concrete to the bottom of a caisson.

Trussed Rafter — (Frame construction) A roof truss which serves to support the roof and ceiling construction.

U

Underlayment — Floor covering of plywood or fiberboard to provide a level surface for carpet or other resilient flooring.

V

Vapor Barrier — A watertight material used to prevent the passage of moisture or water vapor into and through walls.

Vermiculite — Expanded mica used for loose fill insulation and as aggregate in concrete.

Vibrator — A mechanical device used in placing concrete to make certain that it fills all voids. (A screed vibrator is a rotary surface vibrator.)

Vitrified Clay Tile — Ceramic tile baked to become very hard and waterproof.

W

Water Hammer — The noise and force which develops in water supply lines when the flow is suddenly stopped by the rapid closing of a valve or faucet.

Web — The wide vertical part of a beam between the flanges.

Web Member — Secondary members of a truss contained between chords.

Weep Hole — Small holes in masonry veneer walls provided to release water accumulation to the exterior.

Winders — The radiating or wedge-shaped treads at the turn of a stairway.

Y

Y-Branch — A plumbing drainage fitting with a branch or branches extending at an angle of 45 degrees.

Introduction

Originally, buildings evolved from simple shelters used as protection against the elements.

As man learned to use more resources, crude shelters became functional as well as protective. Larger, more permanent structures permitted activities to be moved inside where privacy and security as well as shelter could be found. As simple architecture developed, the design of a structure could be altered to meet the specific purpose it was intended to serve. Thus, a shelter for a single family would be simple; one intended for tribal meetings and ceremonies could be larger and structurally more refined. Buildings also began to reflect the cultural and social values of their owners. For example, a chief or king might have a larger, more ornate structure as a symbol of higher status. Religious structures such as temples became grandiose.

At some point long ago one of these early buildings caught fire. Perhaps a cooking fire ignited the grass roof of a hut or a fire kindled for warmth ignited the animal robes of a simple tent. In any event, some unknown primitive became the first person in history to witness a structural fire.

As those ancient flames died away and the home was reduced to ashes, there must have occurred the dreadful realization that even though these shelter structures were necessary, they were vulnerable to the destructive forces of fire. That sad experience has been repeated countless times.

Modern buildings still serve the fundamental purpose of protection from the elements, but construction technology has carried functional design to a high degree of sophistication. A structure's size and shape reflects its intended use as well as a mix of community tastes, personal desires, and economics. Thus, a hospital is designed to satisfy criteria different from an airplane hangar. An office building, although structurally similar, contains different features than an apartment building. Even the most common of structures, the single-family dwelling, will vary greatly in specific style, size, and details.

In the broad sense, modern buildings provide an environment which enhances and supports the activities carried out within their walls. Large, complex structures such as high-rise hotels, enclosed shopping malls, and convention centers are almost enclosed communities, with thousands of occupants.

A modern building may be viewed as a system providing basic human necessities such as breathable air, water, sanitation, power, communications, and transportation.

Of course, not every building is monumental. But whether it is simple or grand, the ancient realization continues; a fire in a building threatens life and property. For those in the fire service there is the added realization that the firefighter must go into these buildings. The building on fire is the work place of the firefighter.

SCOPE

Fire fighting and its tactics are constantly being challenged due to conditions and environments being presented by new technologies. Architects and builders are constantly creating or improving methods of building design and construction so as to give the consumer the most building for their dollar. However, these "improvements" are in some cases increasing the chances of firefighter injury or death because of the construction component's design. Though the component will be of rated design and construction, its inherent design will promote rapid degradation during a fire.

Bearing this in mind, the firefighter should become knowledgeable and have an understanding of building construction, features of buildings, and how these designs and/or components can serve or hinder the firefighter during suppression operations.

There are many notable incidents where firefighters have been killed or injured because of failing building construction. Firefighters have fallen through roofs, had walls fall onto them, or have become trapped due to a partial collapse of a building. Thus, it is the intent of this manual to help the firefighter become aware of the many construction designs and features that make up the buildings found in a typical first-alarm district.

This manual meets the Fire Officer I objective of knowledge of the characteristics of basic burning materials and their behavior under fire conditions as set forth in the NFPA 1021 *Standard of Professional Qualifications.*

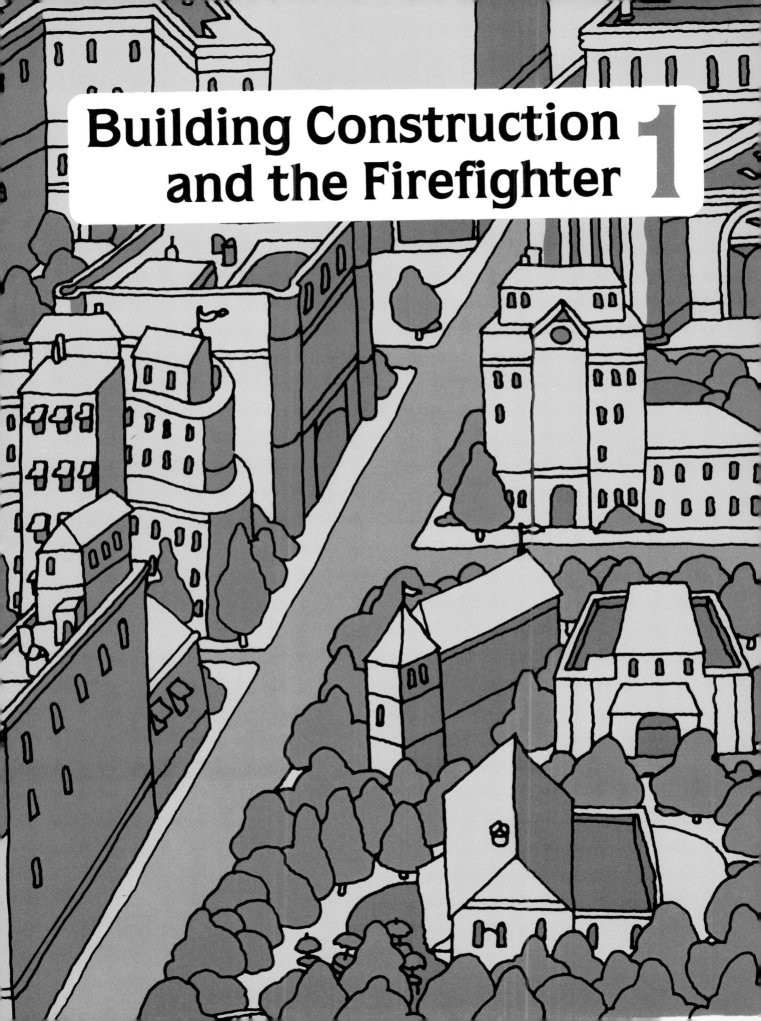

Building Construction and the Firefighter

1

4 BUILDING CONSTRUCTION

Chapter 1
Building Construction and the Firefighter

FIRE BEHAVIOR

A comparison can be made between the occurrence of a fire in a building and an infection in a living body. A living body will respond to repel an infection and may be capable of doing so unassisted, with its own defense mechanisms. If it is not able to do so, extra help is needed from medical science. If the infection is too great, death of the body occurs. In any case, the progression of the sickness through the body will be greatly influenced by the living body itself.

If a building is viewed as a system which provides an environment to enhance or support human activity, then a fire inside is an attack upon the system the way an infection is an attack upon a living body. As in the case of a living body, the course of the fire will be greatly influenced by the way in which the building reacts to it. The efforts of firefighters are analogous to the efforts of doctors to treat a sick patient.

Of course, the building is not the only cause or influence of the behavior of a fire within it. The contents, whether they are simple office furnishings or complex industrial chemicals, will obviously play a role if to do nothing else than contribute fuel. However, even a building as simple as a woodshed affects the behavior of fire within it through the combustibility of the material from which it is built.

Basic features of a building, including its size, materials and methods used in construction, and its configuration will greatly influence fire behavior, occupant safety, and firefighter safety. Ultimately, these considerations determine the tactics to be used by firefighters. For example, modern practice is to rely extensively on total air conditioning to provide a comfortable environment for building occupants. Total control of the heating, ventilation, and air conditioning of buildings has reduced or eliminated the need for windows. Firefighters, therefore, have had to develop techniques to ventilate buildings which have no windows or windows that do not open.

A thorough understanding of the nature of fire is needed in order to understand how building construction relates to fire. Fire is a complex and dynamic chemical process which simultaneously obeys the interacting laws of physics and chemistry. For purposes of discussion, it is useful to differentiate between the ignition of fire and fire behavior.

COMBUSTION

The ignition of fire is basically the fire triangle (or tetrahedron) known to everyone involved in fire fighting. Ignition of a fire requires that some source of thermal energy of sufficient strength be applied to a combustible material in the presence of oxygen. After ignition, the growth of the fire depends on the availability of additional fuel and oxygen.

Ordinarily, the construction of a building has little to do with the initial ignition of the fire, except in those cases where some combustible component of the building catches fire. An example is an overheated stovepipe igniting wood ceiling joists.

If, after initial ignition, additional fuel and oxygen are available, the fire will continue to grow, so attention must shift to the overall behavior of the fire.

When the fire becomes large enough to heat the area in which it is confined, the building itself will begin to influence fire behavior. The spread of fire is often discussed without pointing out that it is more accurate to talk about the spread of heat. In all cases, it is the heated products of combustion that spread ahead of the fire, heating combustible surfaces to their ignition temperatures: the actual combustion follows. The building configuration (the joist spaces, vertical shafts, partition walls, etc.) determines the path that the heated products of combustion will follow, both horizontally and vertically. An example of this is the tendency for a fire to travel up the combustible voids within the wall of a wood frame building. Good fire fighting tactics require that either the fire be stopped in its advance along paths of travel, or that likely travel paths be checked for fire after the main body of fire has been extinguished. One method is to pull down the ceiling in a room adjacent to the room on fire.

STAGES OF FIRE

In addition to building configuration, growth of a fire occurring in a confined space will be determined by the relative amounts of fuel and oxygen available to it. Typical fire behavior can be characterized by certain stages (Figure 1.1). When initial

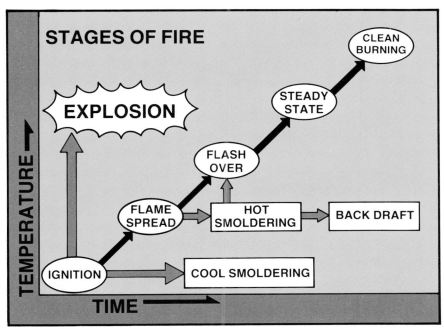

Figure 1.1 Fire grows and spreads dependent on the amount of fuel and oxygen through a series of stages.

ignition occurs there is usually an abundance of oxygen available to the combustion zone from the surrounding air, and it is the fuel and the ignition source which will influence the fire growth. For example, suppose some trash is ignited by sparking from a defective motor (Figure 1.2). As the fire grows it requires more oxygen. If the fire has occurred in a large space (a factory or warehouse), the environment will continue to supply enough oxygen and the fire will progress to the flame spread stage (Figure 1.3). This stage

Figure 1.2 Sparks can ignite a light fuel in the presence of oxygen.

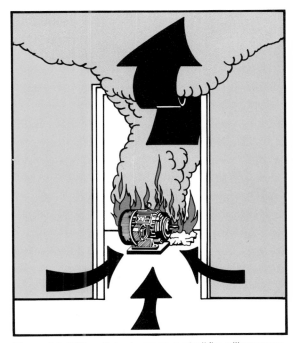

Figure 1.3 With sufficient oxygen, a small fire will progress to the flame producing stage and spread.

will be characterized by a fairly rapid increase in ceiling temperatures to the range of 1000°F (538°C). However, if the fire has occurred in a small, relatively airtight structure, then the development of the fire will be slower, with reduced ceiling temperature. If the supply of available oxygen becomes sufficiently low, the fire may be nearly extinguished (or even totally extinguished). Under these conditions, the amount of heat liberated will not adversely affect the building components. However, enough heat may be generated to vaporize some of the volatile components of combustible contents. This creates the fuel-rich, oxygen-poor atmosphere which is ideal for a "backdraft."

Should additional oxygen suddenly become available, such as by breaking a window with smoldering combustion still present, then the mixture will burn off rapidly, creating a backdraft. The sudden rapid burning will, of course, raise the interior temperatures rapidly. The increase in temperatures causes an increase in the volume of the gases. This increase in volume can create sufficient pressure within the structure to cause structural damage. Buildings are not designed to withstand internal pressures.

BACK DRAFTS CAN CREATE INTERNAL PRESSURES CAPABLE OF STRUCTURAL DAMAGE.

This sequence of events is most likely to occur in tightly constructed, well-insulated buildings with little or no make-up air being supplied by a ventilation system, the very conditions desired for energy conservation. Well-insulated buildings will tend to hold the heat of combustion and keep the interior atmosphere at a hot smoldering stage for a longer period of time.

If the oxygen content in the space is sufficient for increased heat production, the fire could develop to the flashover stage or stage 4 on page 7. Many combustible gases are released when the ambient temperature is between 500°F and 1000°F (260°C and 538°C). They collect at the upper levels because of their low density. Thermal radiation from this hot gas layer and the ceiling heats combustible surfaces throughout the compartment. When sufficient heat has been generated by the fire to raise the gas layer to its ignition temperature, rapid combustion will occur across the ceiling. This is quickly followed by the ignition of the preheated combustible surfaces at lower levels.

If sufficient oxygen continues to be available, the fire will progress to steady-state burning, generating an enormous amount of heat.

Obviously, once flashover has occurred in a compartment, human survival is not possible. However, the temperatures which precede flashover also make human occupancy dangerous without full protective gear. Therefore, warning signs of flashover, such as a thick layer of smoke and hot gas and rising temperatures, will generally force the withdrawal of firefighters prior to actual flashover.

The building materials and configuration will play a significant role in the progression of the fire to flashover. Those building factors which tend to promote flashover include low ceilings, small rooms or compartments, and well-insulated walls or building construction. Since thermal radiation plays an important role in the development of a fire, higher ceilings, which tend to place the hot gas layer farther from the combustion area, tend to delay fire spread and flashover. Large spaces provide more cooler ambient air to dilute the products of combustion and retard the buildup of a hot gas layer at the ceiling. Well-insulated buildings tend to retain the heat being generated by the fire and enhance the reradiation of thermal energy to combustible materials within the building.

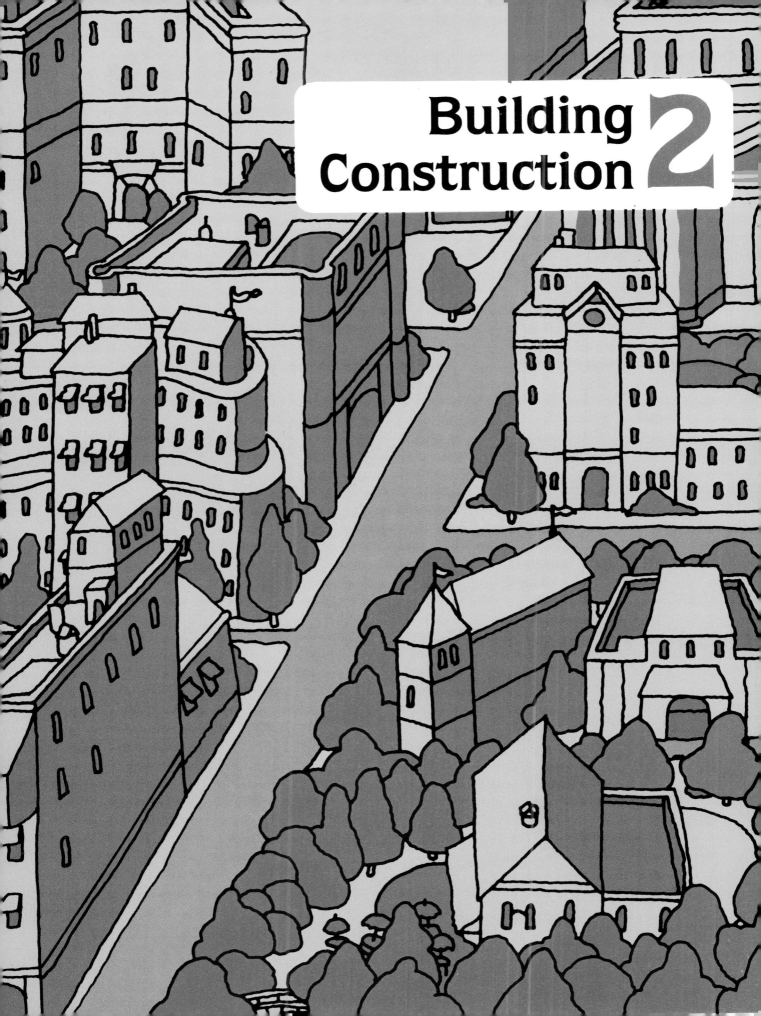

Building Construction 2

Chapter 2
Building Construction

INTRODUCTION

Fire-safe construction can only become a reality when it begins on the drafting board where full consideration can be given to heights, area, exit facilities, interior finish materials, types of structural assemblies, occupancy factors, automatic sprinklers, access for fire fighting, exposure protection, and the limitations imposed by building codes and other authorities.

Unfortunately for those who are responsible for fire suppression, many buildings are designed without regard for fire safety. Designers and engineers are routinely schooled in all the other details of buildings, but the subject of fire protection is included in the curricula of only a handful of colleges. Therefore, a designer's knowledge about fire protection may come only from contact with building or fire officials.

Anytime a fire occurs in a building, a potential exists for each of the following effects:

- The destruction of the contents of the building.
- The death or injury of the building's occupants.
- The ignition of the building itself.
- The partial or total collapse of the building.
- The extension of the fire to other buildings.

The object of effective fire fighting is the elimination or prevention of these undesirable effects.

Of the many unknown factors confronting the firefighter arriving at the scene of a building involved in fire, the one factor that can be changed with the greatest degree of accuracy is the building itself. However, the time to gather the necessary information is before the fire occurs. Many of the buildings that have been the scene of major fires were in existence before many of today's firefighters were born. Yet, in many cases, the building is a complete mystery to those who are responsible for controlling a fire within it. If firefighters are to solve this mystery, they must concern themselves with types of building materials, methods of construction, and how they react under fire conditions. Since few buildings are constructed of a single type of material, firefighters must also concern themselves with the assembly of the different types of materials. Before considering building materials as techniques, an understanding of some of the principles of physics that govern the design, materials, and methods of assembly is needed.

CONSTRUCTION PRINCIPLES

Gravity is the eternal opponent of a building. It is trying to pull the building down twenty-four hours a day, seven days a week. For centuries, designers and builders were primarily concerned with preventing their buildings from collapsing. Unlike gravity, which is a certainty, fire is a probability, and the probability of fire can only be estimated. Building designers have to prove that their buildings are structurally resistant to the known opponent of gravity. The whole philosophy of structural fire pro-

DEAD LOAD — Weight of the structure itself and any equipment permanently attached or built in. Concrete, Steel, Wood, Air Conditioning, Plumbing, etc.

Figure 2.1 Dead loads are the static weights of the building components and attached equipment.

tection would change radically for the better if the builder were also required to prove that a prospective building could be evacuated safely in the event of a fire.

DEFINITION OF LOADS

Specific terms are used to describe the different loads placed on a building. It is important that they be understood and used correctly.

Dead Load

The weight of the structure, structural members, building components, and any other feature that is constant and immobile. For example, the weight of a floor slab that must be supported by columns below it (Figure 2.1).

Live Loads

Loads that may be moved, such as furniture, vehicles, office partitions, people, or warehouse inventory (Figure 2.2).

Static Loads

Loads that are applied slowly and remain nearly constant. For example, the filling of a water tank.

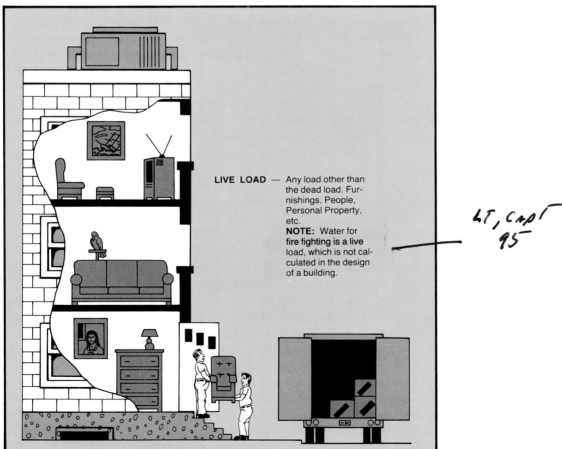

Figure 2.2 Live loads are the dynamic weights that can be moved around within the building.

Impact Loads

Loads that are delivered in a short time with a striking or collision affect (Figure 2.3).

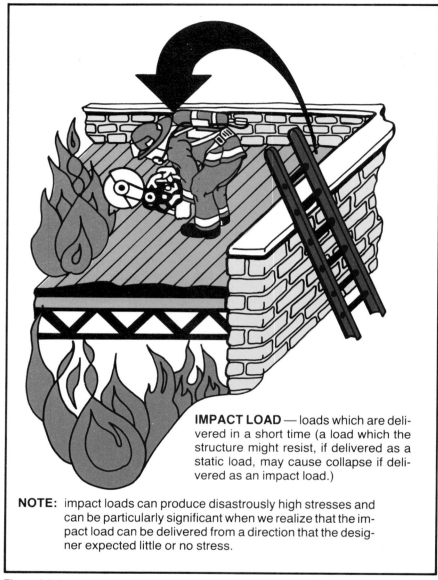

IMPACT LOAD — loads which are delivered in a short time (a load which the structure might resist, if delivered as a static load, may cause collapse if delivered as an impact load.)

NOTE: impact loads can produce disastrously high stresses and can be particularly significant when we realize that the impact load can be delivered from a direction that the designer expected little or no stress.

Figure 2.3 Impact loads unexpectedly strike the bearing surface.

Repeated Loads

Loads that are applied intermittently, such as a rolling bridge crane.

Uniformly Distributed Loads

Loads that are constant over an area.

Concentrated Loads

Loads that are applied over a small contact area.

Wind Load

Force applied to a building or structural member by the wind (Figure 2.4).

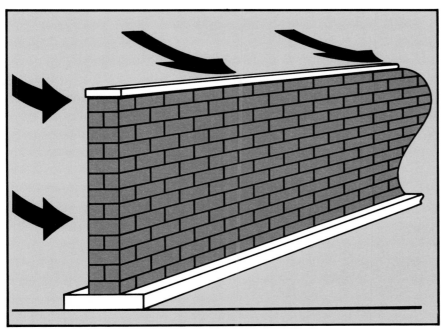

Figure 2.4 Wind loads result from the affects of the wind.

The structural engineer must provide for all of the loads imposed on a structure or it will fail. However, some loads are easier to determine than others. Dead loads are the easiest to design for since they remain constant. For example, structural concrete usually weighs about 150 pounds per cubic foot ($2\ 403\ kg/m^3$). If a concrete beam is to be put in place that measures 6 inches (152 mm) in width by 12 inches (304 mm) in height, the engineer must provide for a dead load of 75 pounds per foot (111.6 kg/m) of beam length. That load will remain constant once the beam is in place. Live loads are more difficult to design for since they are movable and variable. Moving a given load from one point to another will alter the forces on the supporting members. The building codes specify what live loads the engineer must provide for as a function of the intended use of the building. A building code might require that a school classroom be designed for a live load of 40 pounds per square foot ($195\ kg/m^2$) and that a retail store be designed for a live load of 100 pounds per square foot ($488.2\ kg/m^2$).

Live loads are specified as uniformly distributed. Thus, if a column were to support 100 square feet ($9\ m^2$) of floor space in a retail store, the designer would be required to provide for a live load of 10,000 pounds (4 536 kg) uniformly distributed over the supported area. The standards permit a reduction in the total calculated live load when the area is large and unlikely to be totally loaded.

11-88

Individual loads, such as a piece of heavy machinery, will impose loads that are applied over a small area (concentrated). Therefore, a building code might require that, in addition to the uniformly distributed live load, the engineer provide for a specified concentrated load. For example, the floor in a storage area might be required to safely support a load of 2,000 pounds (907 kg) located at any point on that floor.

Static loads are also much easier to design for than impact or repeated loads. Impact loads have an energy associated with the velocity of the moving object which impacts with the structure. This energy must somehow be dissipated when the load comes to rest. Usually, it must be absorbed by the structural members through their deflection or other cushioning action. Repeated loads, depending on how frequently they are applied, can result in a fatigue failure.

Wind loads are probably the most difficult to determine. This is because the actual force exerted on a structure depends on the direction and velocity of the wind, the density of the air, the shape of the building, and the effect of any other buildings nearby. Therefore, building codes specify a minimum wind pressure in pounds per square foot (kg/m^2) that must be provided for. The required wind load will vary with geographic location. In the Midwest 30 pounds per square foot (146.5 kg/m^2) must be designed for, but along the Gulf Coast it may be 45 pounds per square foot (219.7 kg/m^2). Loads imposed by earthquakes and snow are provided for in a similar manner.

Ordinarily, the firefighter is not concerned with wind load. However, it must be remembered that the ability of a wall to resist the force of the wind comes from the horizontal support of the interconnection with the rest of the building. Therefore, extreme caution must be used when operating around buildings under construction where the structural bracing of walls is not complete.

IMPOSITION OF LOADS

Loads are also classified as to the direction in which they are applied to structural members.

Axial Load

An axial load is applied to the center of the cross section of a structural member and perpendicular to that cross section (see Figure 2.5).

Eccentric Load

An eccentric load is perpendicular to the cross section of the structural member but is offset from the center, creating a tendency in the member to bend (Figure 2.6).

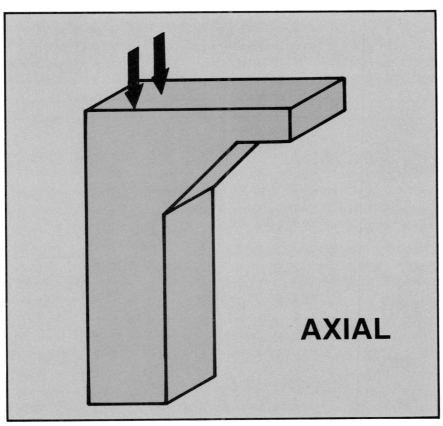

Figure 2.5 Axial loads are applied along the member's axis.

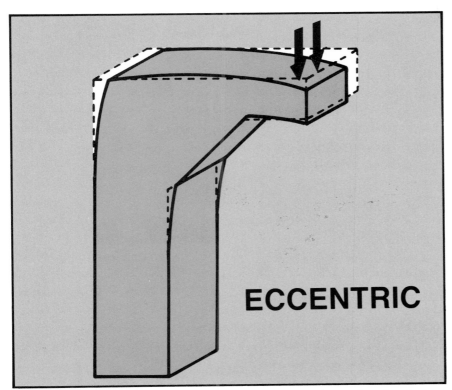

Figure 2.6 Eccentric loads are applied to one side of the cross section creating a bending tendency.

Torsional Load

A torsional load is offset from the center of the cross section and at an angle to or in the same plane as the section, creating a twisting effect in the member (Figure 2.7).

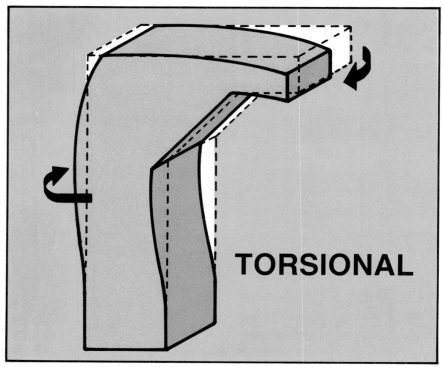

Figure 2.7 Torsional loads are applied at an angle to the cross section creating a twisting tendency.

EFFECT OF LOADS ON MATERIALS

In addition to the manner in which loads are applied externally to structural members, it is important to be aware of the forces within the material that result from the external loads (Figure 2.8).

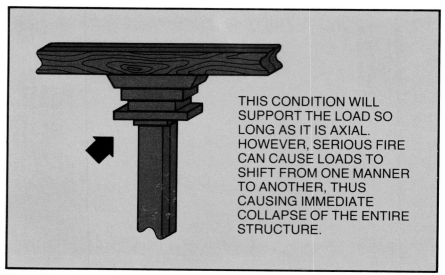

THIS CONDITION WILL SUPPORT THE LOAD SO LONG AS IT IS AXIAL. HOWEVER, SERIOUS FIRE CAN CAUSE LOADS TO SHIFT FROM ONE MANNER TO ANOTHER, THUS CAUSING IMMEDIATE COLLAPSE OF THE ENTIRE STRUCTURE.

Figure 2.8 A change in the external loads, such as under fire conditions, can cause failure.

Compression

A force that is crushing or pushing the mass of the material together (Figure 2.9).

Tension

A force that tends to pull the material apart (Figure 2.10).

Shear

A force that tends to cause adjacent planes in a structural material to slide past one another (Figure 2.11).

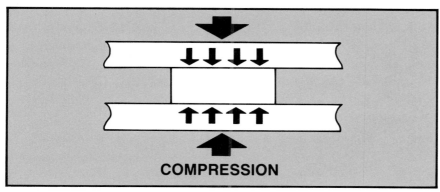

Figure 2.9 Compression forces bring surfaces together.

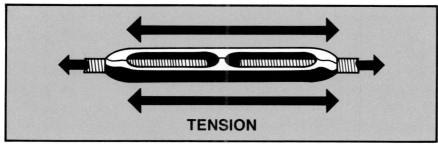

Figure 2.10 Tension forces separate.

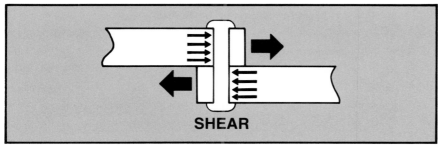

Figure 2.11 Shear forces cause an opposite, but parallel sliding motion.

This concern for the types of loads, the manner in which they are applied, and the resultant internal forces they create arises from the fact that the properties of structural materials vary with respect to direction. The most common example of this is wood. It is a well-known fact that wood will exhibit different properties across the grain than it will in the direction of the grain. Similarly, concrete has a high compressive strength but a relatively low tensile strength. Therefore, when tensile forces are placed on concrete, steel reinforcing bars must be provided to absorb them.

THE SIGNIFICANCE OF STRUCTURAL ENGINEERING TO THE FIREFIGHTER

A scientific knowledge of forces and the properties of construction materials, plus generations of experience, have given the profession of structural engineering a very sophisticated basis for design practices. Structures can be designed specifically for the loads that will be placed upon them. This results in economical use of materials and faster construction, which increases productivity. However, it can make a building more susceptible to structural abuse. When the forces on a structure are different from those for which it was designed, or structural members are modified or deteriorate, failure may occur.

The firefighter must be aware of the following types of structural abuse:

- Subjecting the structure to loads for which it was not designed.

- Structural modifications by unqualified workers or contractors.

- Deterioration.

- The forces associated with the violence of a fire.

It has been pointed out that building codes specify a live load as a function of the intended use of a building. Changing the occupancy of a building, such as converting a school building into a warehouse, can subject the structure to loads greater than it was designed for. Other examples include mounting a heavy air conditioning cooling tower on a roof or using a roof truss to support a machinery hoist.

Structural modification can take the form of removal of portions of bearing walls, cutting openings in bearing walls, and cutting away a portion of a structural member, such as a beam or column, to accommodate a run of pipe.

The life of a building may be 75 to 100 years or more. Over that time the forces of nature (wind, rains, temperature changes, ground settling) can alter the structure. These forces include the erosion of mortar in brick walls, rust and corrosion of exposed metals, and rotting of wooden structural members. Most damage of this type, however, occurs slowly and can be repaired by ordinary building maintenance and repair.

A firefighter encountering instances of structural abuse during a fire inspection should refer the matter to the local building department for structural inspection. If there is evidence of imminent danger, such as sagging or cracking, immediate steps may have to be taken to protect the public.

A fire is a violent assault upon the structural integrity of a building. Virtually all building materials are adversely affected by exposure to the high temperatures of a fire. The differences are in the manner and time required for material to be affected. A slender steel bar joist will soften and yield rather quickly, a reinforced column will remain intact for hours.

From a technical standpoint, the fire will alter the structural systems of a building in many ways. It may burn through combustible structural members, destroying their load-bearing ability. The load will then be transferred to adjacent members which have not been designed to carry it. This can lead to structural failure. A steel beam will initially tend to expand upon exposure to a fire. This imposes horizontal forces upon adjacent columns or walls.

Structural collapses and falling contents impose impact loads on floors. Water from hoses will increase the uniformly distributed load. For example, an accumulation of 4 inches (102 mm) of water will produce a load of approximately 21 pounds per square foot (47.6 kg/m^2). The entire system of loads and supports will deviate from the original design under the influence of the thermal energy released in a fire.

More precise design techniques have enabled the structural engineer to reduce the mass of the individual structural members (especially beams and joists) while retaining the same strength. The widespread use of steel bar joists over the last generation is a result of this more sophisticated analysis. However, lighter weight structural members can complicate fire fighting operations. In general, a structural element which is lighter than the structural element previously used to carry an equivalent load is less fire resistant (unless provided with fire-resistive insulation). The reason for this stems from the mechanics of heat transfer. The transfer of heat into a body is a function of the available surface area. For a given mass, a greater surface area will result in fast heat transfer into the mass. Therefore, a wood truss, which has a relatively high surface to mass, will be adversely affected by fire more quickly than a solid joist. (In this case, by the burning of the truss members.)

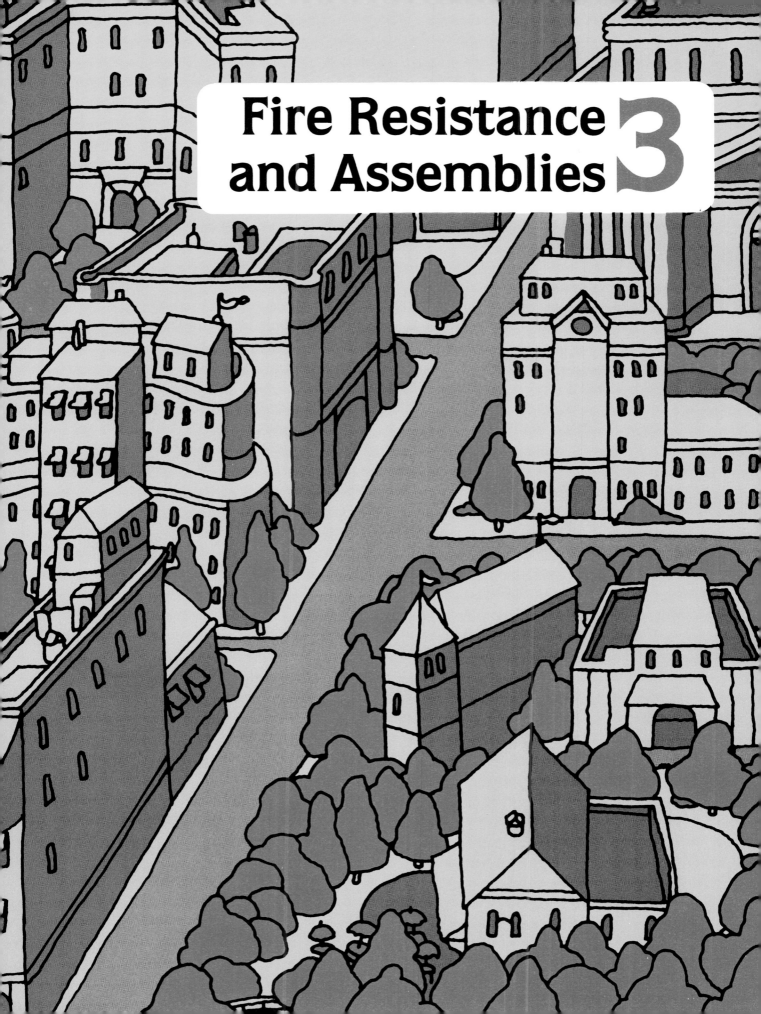

Fire Resistance and Assemblies 3

Chapter 3
Fire Resistance and Assemblies

FIRE RESISTANCE

Since no material is immune to damage by severe and prolonged exposure to fire temperatures, it is necessary to specify the desired fire resistance of materials to be used in a building. This is done on the basis of the expected fire loading or occupancy factors.

Determining Fire Resistance

Fire resistance is a collective property of materials and assemblies. Basically, fire resistance is the ability of a structural assembly to maintain its load-bearing ability under fire conditions. In the case of walls, partitions, and ceilings, it also means the ability of the assembly to act as a barrier to the fire. The fire resistance of a structural component is a function of various properties of the materials used. This includes their combustibility, thermal conductivity, chemical composition, and dimensions.

The fire resistance ratings of materials are determined by fire test procedures simulating fire conditions using the standard time-temperature curve. Fire resistance ratings are given for assemblies of structural elements, such as floors, floor-ceiling assemblies, columns, walls, and partitions. It is necessary to test such assemblies as erected in the field so that values can be assigned which are meaningful to the fire protection engineer or building designer. (A complication is that not all possible assemblies have been tested and listed.)

STANDARD TEST METHODS

In the standard fire test, the furnace temperatures are regulated to conform to the time-temperature curve. The temperature rises to 1,000°F (538°C) in five minutes, to 1,700°F (927°C) at one hour. See Figure 3.1 (on next page). The fire-resistive rating is the period of time that the assembly will perform satisfactorily when exposed to the test fire. Failure of an assembly is determined by one of several criteria. These include failure to support the load, passage of flame through the assembly, and excessive increase in temperature on the unexposed side of an assembly. NFPA 251, *Standard Methods of Fire Tests of Building Construction and Materials* contains specifications of the test procedures.

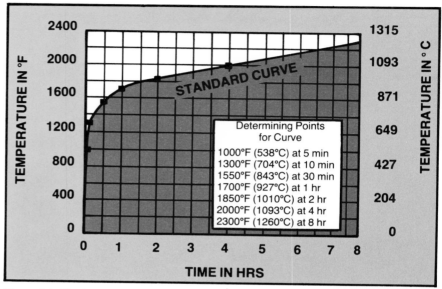

Figure 3.1 The time-temperature curve indicates how rapidly a fire builds up initial heat and then levels as the temperature continues to rise over an extended time period.

Although an assembly may fail at any time, fire resistance is expressed in certain standard intervals, such as 15 min., 30 min., 45 min., 1 hr., 1½ hr., 2 hr., 3 hr., and 4 hr. The fire resistance ratings form the basis upon which types of building construction are recognized in building codes. However, the standard test is a laboratory index of performance for various materials. Not all field conditions can be duplicated in the laboratory. Obviously, size restrictions do not permit testing entire buildings. The field performance may deviate from the laboratory results. This is possible when building components have not been installed with the same craftsmanship used in the laboratory. For example, joints in wall or ceiling assemblies may not have been carefully fitted. Furthermore, some of the lighter weight building materials may not be properly maintained, such as holes being poked through plaster assemblies.

Some laboratories that perform fire-resistive tests are

> Underwriters Laboratories, Inc.
> National Bureau of Standards
> Factory Mutual System
> Forest Products Laboratory
> Portland Cement Association

Probably the best known of these is Underwriters Laboratories, which publishes test results in the "Fire Resistance Directory."

FOUNDATION ASSEMBLIES
Footings

Generally speaking, footings are that part of the foundation assembly which are built below grade and serve as supports for walls, piers, columns, girders, posts, or beams. The footings for any building are designed to transfer and distribute the total weight of the structure and its contents to the ground. The type of footing needed depends upon the size of the structure, the occupancy of the structure, and what type of soil the structure is built upon. The footings may be as simple as those needed for a single-family dwelling (Figure 3.2), or as complex as those needed for a high-rise building (Figure 3.3).

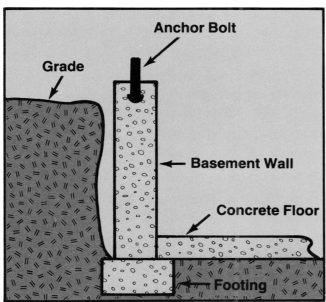

Figure 3.2 Single-family dwelling footing.

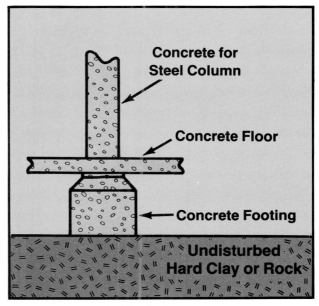

Figure 3.3 High-rise building footing.

Today footings are engineered and designed for a specific structure. However, there are many buildings in existence that were constructed around the turn of the century, and in many localities even earlier, without benefit of engineering and design technology. Further, many structures that were designed for a specific type of occupancy have been modified structurally and the type of occupancy changed many times. These changes have drastically altered the load being carried and distributed by the structural members and/or assemblies.

As an example of this problem, consider a two-story building that was constructed before the turn of the century as a mercantile store, and now is found to be occupied by a furniture company warehouse. Upon careful examination, it will be found that the footings and foundation walls are not of sufficient strength to support the load which has been imposed upon it, and measures have been taken to shore up sagging and failing structural members.

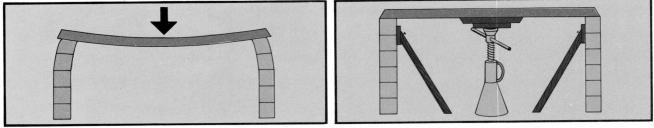

Figure 3.4 Foundation walls may need shoring, and other structural members may need support from jacks when the walls and foundation become overloaded.

Ordinarily, the structural problems of footings and foundations are of little interest to the firefighter. However, they can become of concern to the firefighter in several ways. If a foundation or footings become sufficiently deteriorated or abused, the structure will ultimately collapse. The firefighters are then typically called upon in their role of first responders to perform rescue services in a collapsed structure. Another problem is that shifting or settling supports will naturally alter the forces on the structural members in the upper part of a building. These altered load patterns can hasten structural collapse under fire conditions, as previously discussed on page 22.

Since the footings of a building are not usually visible, especially from the outside under fire conditions, it is impossible for the fire officer to determine the type and condition of footings supporting any structure. Figures 3.5 through 3.7 are examples of types of footings that might be found under any given structure. However, severe deterioration of foundations will usually result in cracked walls, sagging floors, or other signs that can be visually detected by the alert fire officer. See Figure 3.8.

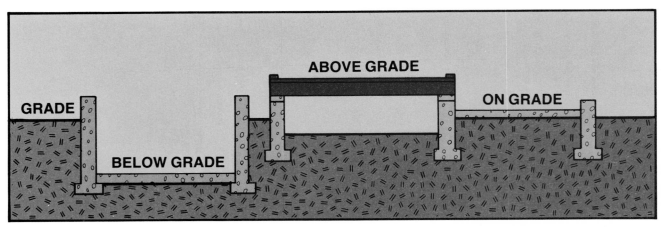

Figure 3.5 Each foundation wall has a footing to support the planned building load.

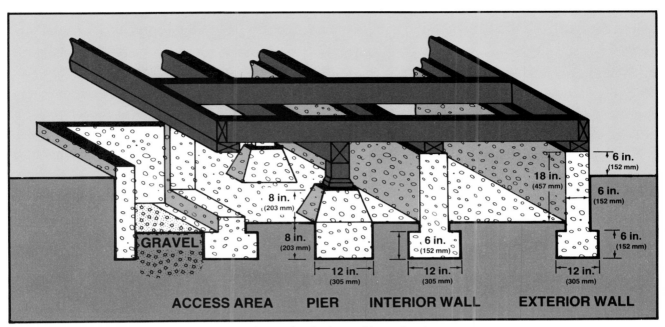

Figure 3.6 The load bearing walls have footings, the pier acts as a footing, and the nonbearing wall has no footing.

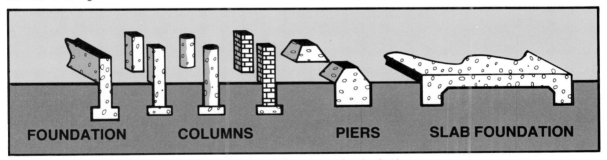

Figure 3.7 Footings are provided under the various foundations except the pier that is a combination footing/foundation.

Figure 3.8 Deteriorating foundations are evidenced by cracked walls, uneven floors, and sagging structural members. *Courtesy of Edward Prendergast.*

Pilings

There are many situations where the weight of a structure is great enough that "ordinary" footings are not sufficient to transmit the load adequately to the ground. This is usually due to soil conditions which are inadequate to support the weight of the structure. This condition requires that pilings be used to transmit the weight of the structure to bedrock or to substrata soil that is of sufficient density to support the load. Pilings may be of wood, steel, or concrete. See Figure 3.9 for details.

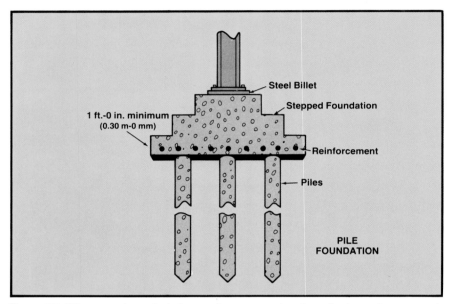

Figure 3.9 Pilings may be a single large column or a series of small columns to an area of ample support.

Floating Foundations

When soil conditions have an extremely low bearing capacity, it may be necessary to use the floating foundation method of supporting a structure. A floating foundation consists of a footing over the entire area or large portion of a building. It is made of very thick, reinforced concrete, and is often used in conjunction with pilings for additional support (Figure 3.10).

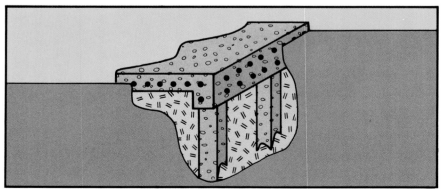

Figure 3.10 A floating foundation combines the footing and foundation into one large reinforced member often further supported by pilings.

Grillage

Sometimes additional foundation support is needed for an existing structure. This may be due to excessive live loads imposed on the structure for which it was not designed, or it may be due to proposed vertical additions to the structure. The structural members may be strong enough to carry and transmit the added load, but the footings may not be able to transmit that load to the ground.

To support this additional load a grillage, or framework of beams, is laid horizontally and crossed with similar beams laid upon them to distribute the vertical loads over secondary footings or underpinnings (Figure 3.11). This practice is not commonly used today due to the cost and difficulty.

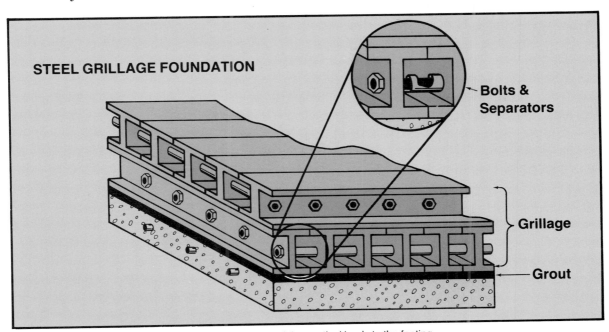

Figure 3.11 A grillage can be designed to assist in transmitting vertical loads to the footing.

Underpinning

It may be necessary to provide additional support to an existing structure to prevent structural damage and possible building collapse. Many structures have been underpinned as a stopgap measure to check imminent collapse. Most often this is done hurriedly, with little or no engineering technology used. It may well be done by the building's owner or occupant.

FOUNDATION WALLS

Foundation walls are considered to be those walls, piers, and columns below grade level which serve as the supporting members for the structure and transmit that weight to the footings.

Foundation walls are normally constructed of concrete, stone, brick, or concrete block. However, wood foundations are

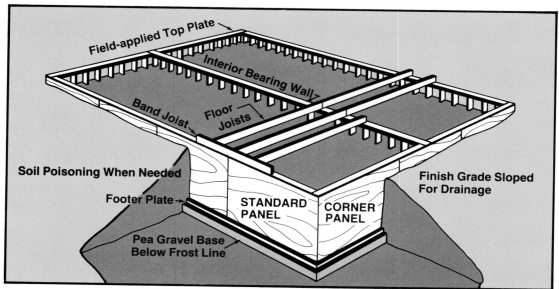

Figure 3.12 Although below grade, wood foundation walls are being used in some residential and small commercial applications.

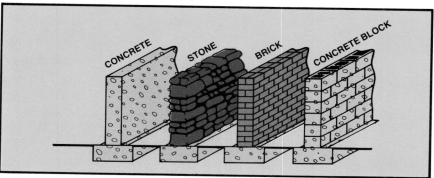

Figure 3.13 Foundation walls transmit the weight of the building to the footings.

now being used in residential and commercial applications. (Figures 3.12 and 3.13).

Piers and Columns

Piers and columns are vertical members which are used to support the floor systems. Materials used are usually concrete, brick, steel, or wood (Figure 3.7). Piers and/or columns may be used in place of walls, or they may be used in conjunction with the foundation wall, and provide only intermediate support for girders and beams (Figure 3.6).

Concrete Foundation Walls

Present-day technology and engineering determine the physical size and materials to be used in the construction of foundation walls. Poured in place concrete has relatively good compressive strength. However, its resistance to tension forces is much less than that of other materials. Therefore, steel is used in conjunction with concrete in foundation walls to greatly increase its strength and resistance to lateral forces. The steel used for reinforcing concrete is usually in the form of rods ranging in diameter from ⅜ inch to 1⅜ inch (10 mm to 35 mm). The amount of reinforcing steel and the exact configuration of rod placement depends upon the engineering requirements that have been predetermined for the load requirements.

Concrete has good resistance to fire and heat. However, spalling of concrete can occur when it is exposed to extremely high temperatures.

Effects of erosion in concrete are probably more visible to the layman than in other types of building materials.

Many concrete foundation walls will develop visible cracks for varying reasons. They usually do not significantly affect the ability of the wall assembly to support or distribute the load which it is carrying. However, when inspecting a structure for stability, any change in size or extension of cracks or fissures should be given close attention. Any vertical or horizontal misalignment along the length of a crack in a foundation wall indicates a movement or shift in the structure, which may mean a change in imposition of loads on structural members. (See section on loads, imposition of loads, and load transmission.) It should be remembered that foundation walls are carrying and transmitting the entire load of a structure and its contents to the footings.

Masonry Foundation Walls

These types of walls are used extensively for foundation walls in all types of buildings. They are usually constructed of smaller component members bonded together by mortar to form a continuous interlocking assembly. The individual component members may be concrete block, stone, brick, or clay tile. In general, a masonry wall has load-carrying capabilities similar to those of concrete. However, due to the bonding process used in all masonry construction, there is an inherent weakness in resisting lateral forces. Masonry walls do not have the shear strength of solid concrete. Any failure of a masonry wall usually occurs along the horizontal bonding plane between component members of the assembly. Reinforcement of masonry walls can be done in several different ways (Figures 3.14 and 3.15).

The strength of masonry walls depends greatly upon the configuration of the component member when they are assembled. The ability of the entire wall assembly to support and transmit the load is determined by the capability of any single component member to transmit its load to adjoining members.

Clay structural tile and brick were once used in wall construction, but larger concrete blocks have replaced them due to labor costs and speed of erection. However, clay tile and brick can usually withstand higher temperatures than concrete block.

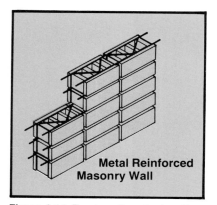

Figure 3.14 To strengthen the bonding plane, masonry foundation walls may be reinforced with metal rods.

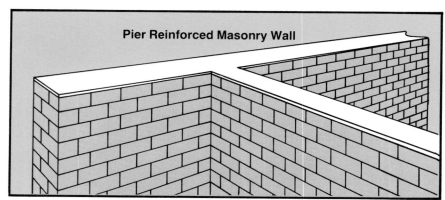

Figure 3.15 Masonry walls can also be strengthened against lateral forces by adding reinforcing piers.

Stone Foundation Walls

These walls were once used extensively or almost exclusively. Today, new stone foundation walls are nearly nonexistent. It should be pointed out that while stone was mentioned as a possible component member in a masonry wall assembly, the individual stones used would weigh up to 100 pounds (45.3 kg). However, the individual component of a "stone foundation" could weigh from several hundred pounds (kilograms) to several hundred tons (metric tons). A distinguishing factor of stone foundations is that they were often constructed without using any bonding mortar or cement. Most of the stones used in this type of foundation assembly were limestone, granite, or flagstone, which were carefully quarried and transported to the building site. They were then meticulously assembled to form an amazingly tight-fitting and strong foundation. Due to the size and weight of stones used, and the degree of difficulty in handling and transporting them, this type construction was usually found where large deposits of limestone and granite were readily accessible.

FLOORS AND CEILINGS

Floor Assemblies

A floor provides the supporting surface for the contents of a building (loads). Where the floor rests directly on grade, it is usually concrete and of little concern to the firefighter. In the countless buildings where the floor itself is supported by some other part of the building, it is a basic aspect of the building's structure and of primary interest to fire personnel.

To the firefighter, a floor has at least three characteristics which must be taken into account during interior fire fighting operations.

- The structural integrity of the floor under fire conditions (fire resistance)

- Its ability to block the vertical spread of fire

- The ease with which it can be breached for purposes of drainage

All floor systems share common functional components. These are the floor deck, supporting joists, and girders. The total load is supported on columns or bearing walls (Figure 3.16).

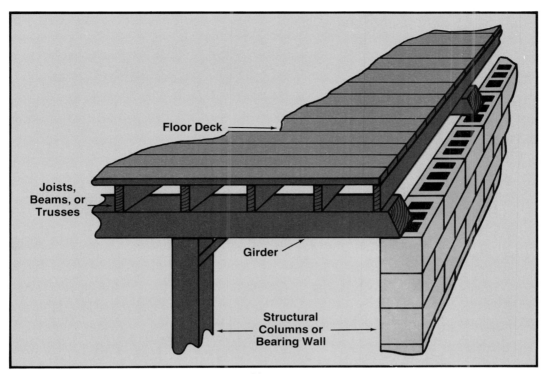

Figure 3.16 The basic components of a floor assembly.

The individual components will vary depending on such factors as the intended use of the building, the loads, span lengths, and aesthetics. Thus, the floor deck may be plywood or plank, or it may be concrete. The joists may be wood, concrete, steel, or wood or steel trusses. The girders can be heavy wood beams as used in mill construction, steel beams, or concrete (Figure 3.17). In some applications such as "flat slab" concrete floors or where spans are short, girders may not be necessary (Figure 3.18).

Figure 3.17 A concrete floor assembly with the flooring, beams, and girders all of poured concrete. *Courtesy of Edward Prendergast.*

Figure 3.18 The flat slab concrete construction does not require beams or girders to support the short spans. *Courtesy of Edward Prendergast.*

CONCRETE FLOORS

One reason concrete is widely used as a building material is that it is extremely versatile. Until the hardening process has reached a certain point, concrete is a plastic material and can be formed into almost any desired shape. It can be formed into concave and convex shapes for domes, roofs, and arches. Concrete itself is noncombustible and is often used to "fireproof" (insulate) steel.

In concrete construction two terms are used: plain concrete and reinforced concrete. To understand the difference, we must go back to basic considerations. It has been previously pointed out that concrete, which is a mixture of portland cement, fine aggregate (sand), and coarse aggregate (stone), has a high compressive strength but little tensile strength. The compressive strength of concrete will average around 6,000 psi (41 370 kPa), but the tensile strength will average around 500 psi (31 448 kPa). To give it the tensile strength required for most building needs, steel rods are embedded in the concrete where tensile forces occur. This is reinforced concrete (Figures 3.19 and 3.20).

Plain concrete, without reinforcing rods, can be used only where the imposed load is compressive. Sidewalks, cellars, and foundation walls for small residences are often of plain concrete because the loads imposed are usually small and compressive rather than tensile.

Figure 3.19 Reinforced concrete has a grid of steel rods to increase the tensile strength.

Figure 3.20 Reinforced concrete is used extensively for high-rise construction, parking ramps, and other structures that have high tensile strength requirements. *Courtesy of Edward Prendergast.*

Poured in Place

Concrete can also be "poured in place" or "precast." Poured-in-place concrete is placed in forms at the construction site before hardening. The forms give the concrete the final desired shape. Reinforcing steel is placed in the forms prior to placing the concrete. Since structural concrete of the proper quality is somewhat stiff, it must be forced around the reinforcing steel to ensure proper bonding to the steel and to eliminate voids. After the concrete hardens the forms are removed. When placing a concrete floor, it is not unusual to have electrical conduit and other services also put into the forms.

Precast

Precast concrete, as the name implies, is cast and allowed to harden prior to being placed. The hardened concrete components are then transported to the construction site for installation. Precast structural components are available in a wide variety of shapes and sizes, including slabs, beams, and T-beams (Figure 3.21).

Precast slabs used for flooring are usually hollow to conserve weight and to allow for utility services (Figure 3.22). These slabs will usually be topped with a layer of concrete, tile, or other surface finish (Figure 3.23).

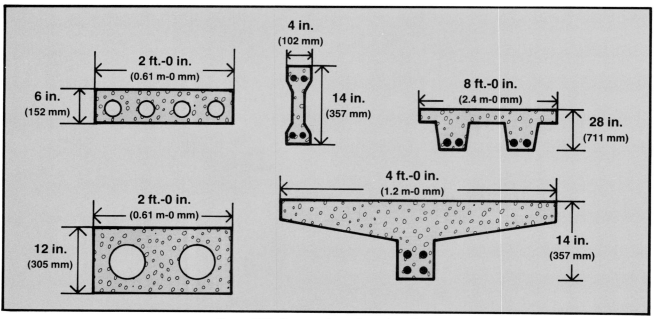

Figure 3.21 Cross sections of a variety of structural members that are available in precast concrete.

Figure 3.22 A hollow precast floor being installed. *Courtesy of Edward Prendergast.*

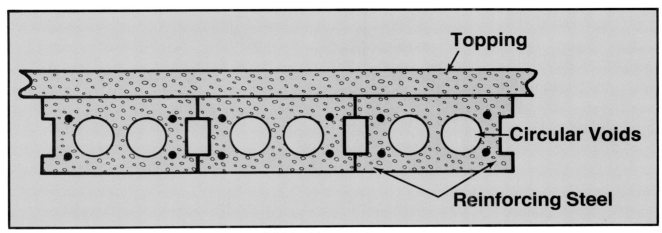

Figure 3.23 Precast floors are provided with a surface finish for tightness and appearance.

Prestressed

The load-carrying ability of concrete may also be increased through prestressing. Since concrete is weak in tension, one method of strengthening it is to impose a compressive force on the concrete prior to the application of loads. When the structural loads are applied, the resulting tensile forces are compensated for by the existing compressive forces.

The prestressing technique allows concrete members to be thinner than ordinary reinforced concrete and is especially useful where long spans are involved or where the dead load of the concrete must be minimized.

PRETENSION AND POSTTENSION

Prestressing can be accomplished in one of two ways: pretensioning and posttensioning. Pretensioned concrete is hardened after the stress is applied; posttensioned concrete is hardened before. In pretensioned concrete, high-strength steel strands are stretched between abutments. The concrete is then placed in forms that are built around the strands. As the concrete sets, it bonds to the tensioned steel. When the concrete hardens, the tensioned strands are released from the abutments. This prestresses the concrete, putting it under compression, thus creating a built-in resistance to loads that produce tensile stresses.

In posttensioned concrete the concrete is placed around the reinforcing steel and allowed to harden. Then a tensile force is applied to the reinforcing steel by means of jacks reacting against the hardened concrete. This results in a compressive force being applied to the concrete, as in pretensioning. In posttensioned concrete, provision must be made to allow the steel to slip within the concrete. This is accomplished by wrapping or placing the steel in conduit. Posttensioning can be accomplished at the construction site. This is useful where the members are large or too heavy for easy transportation to the site. See Figure 3.24 on next page.

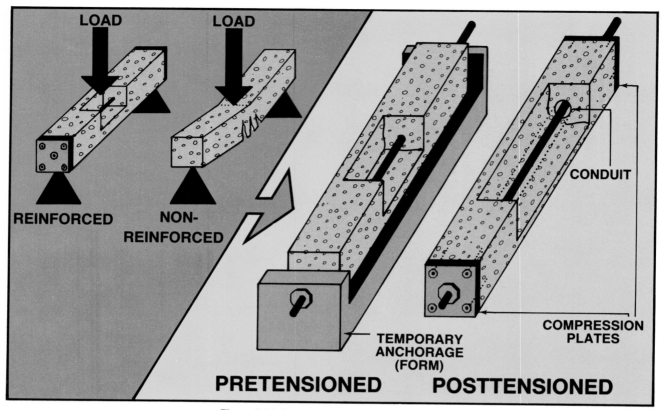

Figure 3.24 Prestressed concrete can be developed by pretensioning or posttensioning.

The idea of prestressing is not new. The first patent was taken out in 1888, but it was not until 1951 that the first major prestressed structure was erected in the United States.

Concrete floors have good fire resistance. For reinforced concrete to fail under fire conditions, it would be necessary for the reinforcing steel to become sufficiently hot to yield or lose its tensile strength. The steel, however, is embedded in the concrete that provides insulation against the fire. This can happen if the steel becomes exposed to the fire. If the concrete surrounding the steel is exposed to a severe fire for a long period of time, it will begin to spall.

Spalling occurs when excess moisture trapped within the cement of the concrete expands. This expansion of the moisture results in tensile forces within the concrete, causing it to break apart, and expose the steel to the fire.

A concrete floor can also fail if the reinforcing steel is cut. Ordinary reinforced concrete will not undergo catastrophic failure if only one or two reinforcing bars are cut, due to the safety factor usually incorporated in structural design. However, cutting the steel in prestressed concrete could result in sudden failure. Therefore, as a general rule, cutting reinforced steel with power tools should be avoided except when a rescue operation makes it necessary.

STEEL-SUPPORTED FLOORS

Concrete is fire-resistive and versatile. However, it is difficult to penetrate for utilities and is not as strong as steel. Therefore, floor systems have been developed that use steel beams to support concrete floors. These systems can be either concrete slabs supported by steel beams, or a concrete slab on a steel deck supported by steel bar joists (Figures 3.25 and 3.26a and b). Notice that in the concrete floor slab the reinforcing steel is placed in the top of the slab over the supporting beams. This is because tensile stresses occur in the top of the slab at these points.

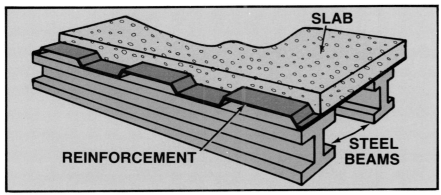

Figure 3.25 Concrete slabs reinforced and supported by steel beams make a strong floor assembly.

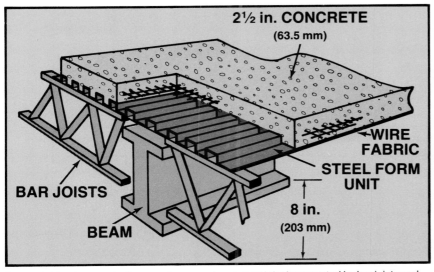

Figure 3.26a A concrete floor can be poured on a metal deck supported by bar joists and steel beams.

Figure 3.26b Underside of concrete floor or steel deck with steel supporting members. *Courtesy of Edward Prendergast.*

When bar joists are used to support a concrete floor, small bars placed 12 inches (305 mm) on center both ways or welded wire fabric will be used for reinforcement. The concrete is placed on a metal lath or deck which acts as the form. A concrete floor deck can also be supported on a lightweight cellular steel panel. The steel cells that protrude into the concrete provide some of the needed tensile strength. These floor systems can be used to span distances of 20 feet (6.2 m) and are used where high live loads are not anticipated.

In general, when a structural beam supports a load, two types of forces occur *within* the beam. These are a bending moment and a shear force (Figure 3.27). To be structurally sound, a beam must have the strength to resist both forces (Figure 3.28).

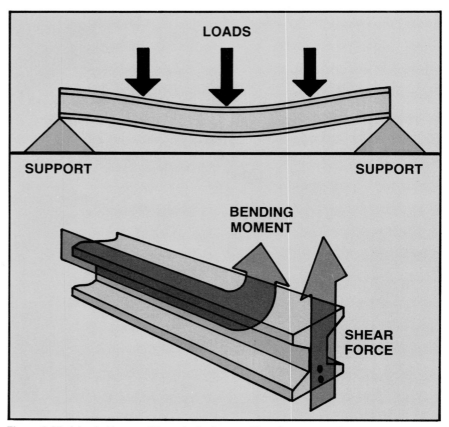

Figure 3.27 A loaded beam will have both shear and bending moment forces working in it.

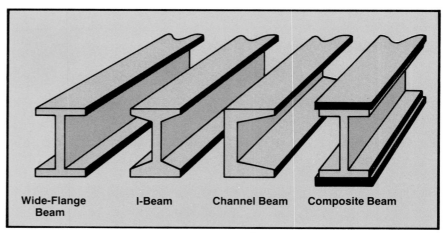

Figure 3.28 Various beam configurations have been designed to resist the forces that affect the beam.

Research engineers have determined that the stresses in a beam resulting from the bending moment are greatest at the top and bottom of the beam. Or, to state it differently, the center section of the beam carries a small portion of the bending moment. The shear force, however, is distributed across the vertical cross section of the beam. This knowledge has resulted in the design of structural members such as I-beams, bar joists, and trusses that have the same strength but are much lighter than a beam with a simple rectangular cross section.

Wide flange I-beams can vary in height from 4 inches to 36 inches (101 mm to 915 mm) and will weigh from 7.5 pounds per foot to 730 pounds per foot of length (3.4 kg to 331 kg per .3 meter).

It should be pointed out that the stresses that occur within a structural member are a result of the shape of the member and the loads that are applied to it. They are independent of the material. Once the stresses have been determined through engineering analysis, it is the responsibility of the engineer to select a material which is strong enough (or change the shape) so that failure does not occur.

To a structural engineer, failure does not mean structural collapse only. Failure can also mean such unspectacular but undesirable effects as excessive deflection, squeaks (in wood floors), or cracking.

The most common structural steel used has the designation ASTM A36. It is a carbon steel with a carbon content between .25 percent and .29 percent. Structural steel weighs approximately 490 pounds per cubic foot (7 84369 kg/m^3). The properties of steel can be altered by varying the carbon content or by the introduction of alloys. Other structural steels include ASTM A242 which is a low-alloy steel stronger than A36. A440 is a high-strength steel, stronger than A242. A441 is a manganese vanadium steel. The alloying elements that can be added to structural steel include silicon, manganese, copper, nickel, and vanadium.

WOOD FLOORING

As a construction material, the physical strength of wood is more variable than that of steel or concrete and requires higher factors of safety in design. However, it is a material that can be easily worked in the field (cut, nailed, drilled). Furthermore, it has adequate strength for many applications. For these reasons, wood has been and continues to be used for many structural purposes.

Wood flooring has the fundamental disadvantage of being combustible. It will add 5 pounds per square foot to 15 pounds per square foot (24 kg/m^2 to 73 kg/m^2) to the structural fire load. One practical consequence of this is that vacant buildings with wood floors and structural components can have serious fires.

It is virtually impossible to have a significant fire in a vacant noncombustible or fire-resistive building.

When exposed to fire, wood will ultimately fail structurally. Whether this failure occurs rapidly or slowly, and whether it occurs with or without warning, are matters of primary concern to the firefighter.

During the last century, prior to the development of rolled steel beams, multistory buildings were constructed with load-bearing masonry exterior walls and wood interior structural members. When heavy floor loads were anticipated, the wood components had to be of large enough dimensions to have adequate load-carrying ability. This gave rise to a type of heavy timber construction referred to as mill construction. The use of the word "mill" is derived from the fact that this type of construction was frequently used to support the machinery in New England textile mills.

In mill construction, wood columns are usually 8 inches by 8 inches (203 mm by 203 mm) minimum. Beams and girders are 6 inches by 10 inches (152 mm by 254 mm) minimum. The wood flooring is usually 3-inch (76 mm) plank with 1-inch (25 mm) top flooring or laminated 4-inch (101 mm) wood planks laid on edge (Figure 3.29). This construction technique results in a structure

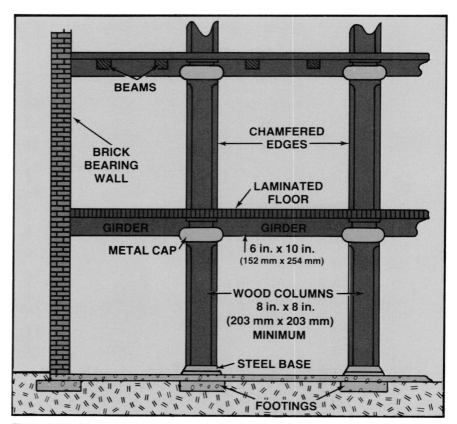

Figure 3.29 Mill construction uses heavy timber beams and girders to support a heavy plank or laminated wood floor.

which, although combustible, burns relatively slowly and retains its structural integrity for a useful period of time (½ hour to 1 hour depending on exact dimensions and severity of fire exposure).

In mill construction, the edges of the columns were chamfered. This was done to make them harder to ignite, thus increasing their fire resistance.

Wood flooring is also used in occupancies where the floors do not require massive structural members. Joists 2 inches by 8 inches (51 mm by 203 mm) are commonly used with various types of decking, including plywood (Figure 3.30). As the dimensions of wood members become smaller, their fire resistance drops rapidly. This is because the burning rate is a function of the ratio of the surface area to the mass of a member, and smaller members have greater surface area-to-mass ratios.

Various types of assembled beams can be used instead of solid beams. These include trusses, laminated beams, and box beams.

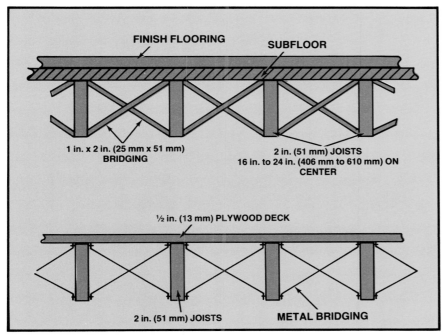

Figure 3.30 Ordinary construction uses bridge-supported joists with a subfloor or plywood deck covered with a finish flooring.

Trusses

Wood trusses have been developed to support floors. They consist of a flat top chord, a bottom chord, and web members. The web members can be either wood or metal (Figure 3.31). The component members of wood have considerably less mass than corresponding solid joists or beams. Consequently, they will burn and fail more quickly than solid wood members. Unless otherwise protected, wood floor trusses cannot be relied upon for any useful degree of fire resistance.

Figure 3.31 Wooden floor trusses are being used in newer construction. Note plywood deck and installation of utilities. *Courtesy of Edward Prendergast.*

Wood floor trusses are usually spaced 24 inches (610 mm) on center. The openings between the web members permit convenient installation of plumbing and electrical lines.

Laminated Beams

It is possible to construct a wood beam from flat pieces of lumber laid on their sides. Such a technique permits smaller individual sizes of lumber to be used, with better control of moisture. However, a beam constructed from such laminations will have poor resistance to bending unless a means is provided to prevent the individual laminations from sliding on each other. The most common method is to glue the laminations together under pressure in a factory. Such beams are called "glulams."

Individual laminations are usually 1½ inches (38 mm) thick. Widths of laminated beams vary from 2½ inches to 14¼ inches (64 mm to 362 mm). Laminated beams can be constructed in lengths up to 100 feet (31 m). To obtain the necessary length, scarf joints are used to join the ends of individual laminations (Figure 3.32).

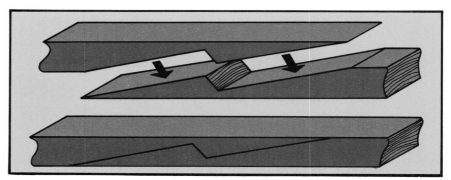

Figure 3.32 Laminated beams use scarf joints to join shorter pieces in the lamination.

Box Beams

Box beams are composite beams made with vertical plywood webs. The plywood webs are attached to lumber upper and lower flanges. The flanges will be separated by stiffeners at regular intervals to evenly distribute the load. This technique provides a high strength-to-weight ratio but is not frequently used (Figure 3.33).

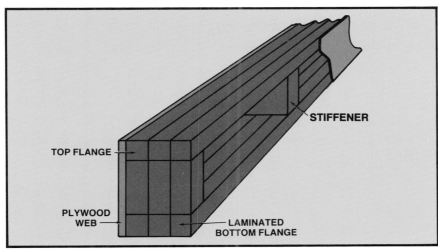

Figure 3.33 Box beams are lightweight and strong.

Plywood Beams

Beams can be fabricated using plywood of various thicknesses for the vertical web portion of the beam. Plywood beams can be made in either an I beam cross section or a box beam cross section (Figure 3.34). They can be fabricated with nails, bolts, glue, or combinations of these.

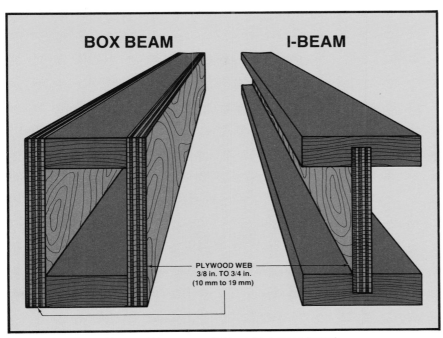

Figure 3.34 Plywood box and I-beams use lightweight plywood for web.

By varying the total height of the beam, the dimensions of the flange lumber, the thickness of the plywood, and by providing stiffeners, plywood beams can be used for spans of up to 100 feet (30.5 m) for light loads. However, most common applications such as in flooring are for spans considerably shorter than this. Engineering data in design tables exists to facilitate the design of plywood box beams with vertical dimensions of up to 48 inches (1 219 mm). A plywood beam provides stiffness with light weight. However, the thinness of the plywood and its combustibility render them highy susceptible to rapid failure under fire conditions.

CEILINGS

A ceiling of itself has no structural role. It supports only its own weight and the weight of attached fixtures. However, a ceiling affects structural fires in two ways. First, a ceiling can be constructed to provide fire resistance to a floor assembly which otherwise has little or no fire resistance. Second, the space between a ceiling and a floor can become a path for the communication of fire.

Fire Resistance of Floor and Ceilings

Floor and ceiling assemblies are rated as a unit when testing for fire resistance. As has been noted, the fire resistance of a floor alone can vary from that of reinforced concrete, which has good fire resistance, to that of plywood decking, which has none. There are a large number of floor-ceiling systems that can be used to provide a fire resistance of from one to four hours. They are published in various reference books available to the firefighter. Possibly, the best of these is the *Fire Resistance Directory*, published by Underwriters Laboratories, Inc.

When a ceiling is exposed to fire, it must protect the floor supports above it from very high temperatures. The steel frames which support the ceiling material can distort, allowing tiles to fall out, or the fire can create pressures which push out tiles.

Various ceiling materials are available which will essentially shield a floor from a fire below and provide a measurable fire resistance to the floor-ceiling combination. A common example of this is the use of a suspended ceiling of mineral tile beneath a floor that is supported on unprotected steel joists (Figure 3.35).

The effectiveness of the floor-ceiling assembly depends as much on adherence to proper installation requirements as on the materials used. Deviation could result in a system that will not stand up properly to the heat of a fire. For example, metal runners used to support ceiling tiles are designed to stay in place when they expand from heat. Also, some ceiling tiles utilize hold-down clips so that they do not drop out under fire conditions.

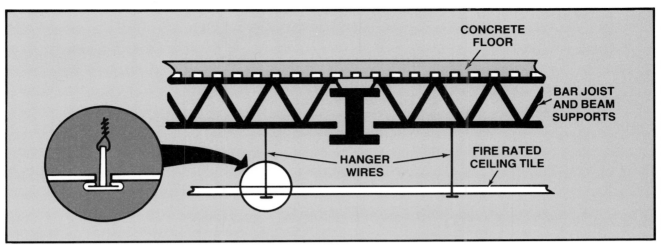

Figure 3.35 A suspended ceiling protects the floor supporting members above it.

SUSPENDED CEILINGS

The suspended ceilings used for fire resistance are a type of ceiling known as membrane ceilings. Since they act as barriers to the heat of a fire they must be complete. Some ceilings utilize a lightweight "drop in" tile supported by a T-bar suspension system. These lightweight tiles can be easily removed for maintenance or other purposes. If the tiles are not replaced, the opening creates a breach of the thermal barrier. Under fire conditions, the heated gases can then enter the space between the ceiling and the floor and weaken the floor supports (Figure 3.36).

Figure 3.36 An opening in a suspended ceiling allows heat to spread laterally and also weaken the directly exposed floor supporting members.

The use of suspended membrane ceilings permits a space to exist between the floor and ceiling. This space can be used for such utilities as electric lighting and ventilation. Where light fixtures or other appliances extend through the membrane ceiling, provision is generally required for protection of the opening. This is usually done by surrounding the light fixture with suitable insulating materials (such as mineral wool batts or gypsum wallboard) and providing dampers within the ventilation ducts (Figure 3.37). In some assemblies, however, insulating material may not be required for a properly tested lighting fixture of limited area. Furthermore, tests have been made which indicate that a limited number of small openings have little effect on fire resistance. An example of this would be an individual electrical outlet with an area less than 16 square inches (1 032 mm^2).

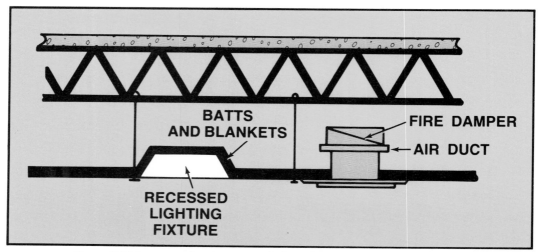

Figure 3.37 Large openings in a suspended membrane ceiling must be protected to restrict fire spread and heat damage.

Fire-resistive membrane ceilings can take several forms. The suspended ceilings discussed so far can make use of mineral tiles, gypsum wallboard, or plaster on suitable lathing. Also, the ceiling material can either be suspended as previously described or attached directly to the bottom of the joists (Figures 3.38 and 3.39). Ceilings can be used to provide fire resistance for combustible as well as noncombustible flooring. For example, a one-hour fire-resistive floor and ceiling assembly can be produced by attaching a ⅝-inch (15.8 mm) layer of type X gypsum wallboard to the bottom of wood joists supporting a 1-inch (25 mm) wood flooring and 1-inch (25 mm) subflooring.

Communications of Fire in Ceilings

In addition to being useful for increasing structural fire resistance, ceilings can provide a more attractive interior finish. Suspended ceilings can conceal plumbing, wiring, and ventilation on ductwork. Also, a new (or false) ceiling can be installed beneath

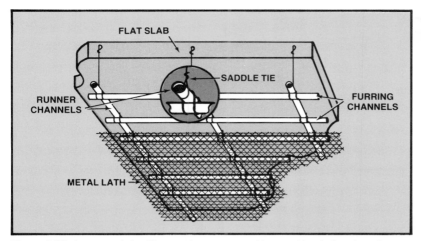

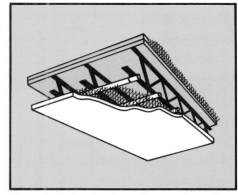

Figure 3.39 Another form of suspended ceiling attaches directly to the bottom of the joists.

Figure 3.38 A suspended ceiling can be constructed by attaching it directly to the underside of the floor.

an existing ceiling when an occupant desires a new decor. The older ceiling can then be left in place.

Building codes usually specify a minimum floor-to-ceiling height for certain occupancies; so new ceiling can be installed only where an adequate height will exist between the floor and the new ceiling. This technique is frequently used in older buildings which often have high ceilings. The space between a new and an old ceiling can vary from almost zero (where the new ceiling is attached directly to the old) to several feet (Figure 3.40).

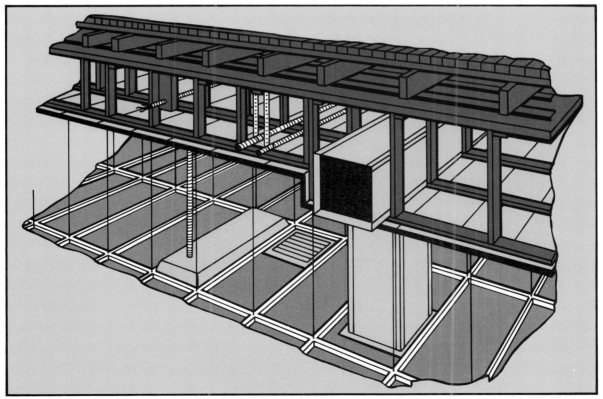

Figure 3.40 New dead spaces are created in buildings when a second dropped ceiling is installed below the original.

The concealed space between a ceiling and a floor can become a significant path for the communication of fire. The extent and speed of communication will depend on the size of the void and the combustibility of the construction materials. For example, if a wire lath and plaster ceiling has been applied directly to the underside of a concrete floor, little problem exists. There is essentially no fuel in the concealed space and any fire gases which penetrate into the space will be blocked by the configuration of the concrete joists. However, where the flooring is wood and the ceiling is suspended some distance beneath the joists, the fire can communicate rapidly above the ceiling. This situation constitutes a classic tactical problem for the firefighters.

A fire can develop in a ceiling space in one of two ways: it can penetrate into the space from below, or it can originate in the ceiling space from such causes as defective electrical equipment. Since the space above a ceiling is hidden from view, the extent of fire spread through the space is difficult to determine. The fire can communicate over the heads of firefighters and into the story or attic above the floor of origin. Ultimately, of course, this will create a peril to firefighters and increase property damage. Good fire fighting tactics call for the ceiling spaces to be opened to halt the spread of fire whenever the slightest indication exists that the fire has entered the space. Such indications include ceiling surfaces that are hot to the touch some distance from the observable fire, discoloration of ceiling tiles, or simply a large volume of fire in a room.

Building codes will sometimes require that large ceiling spaces be subdivided by noncombustible or fire-resistive partitions. This is frequently required where a common ceiling space extends over several rooms or corridors. Fire-resistive partition walls can be made to extend past the ceiling directly to the underside of the floor slab.

In sprinklered buildings of combustible construction, sprinklers are required above the ceiling when the space exceeds six inches (152 mm). See NFPA Standard 13.

Hidden Spaces

A hidden space in a building is an area that is enclosed by structural components, interior finishes, or a combination of these systems. Hidden spaces include false roofs, ceilings, soffits, plenums, utility chases, shafts, or any other areas that are not accessible for occupant use. Hidden spaces provide avenues for air circulation or insulation, the distribution of building utilities, and unfortunately, avenues for fire and smoke extension. A fire can free burn between a wall and extend into an attic without being noticed because of hidden spaces.

Hidden spaces provide avenues for the communication of smoke and fire through both man-made and natural forces. If a fire occurs in a plenum and ventilation fans are not shut down, the fire, heat, and smoke can travel through this space, extend into rooms through HVAC openings, and result in considerable damage. Because of this, firefighters need to be aware of hidden spaces and must check these areas for fire and smoke during the initial stages of the fire attack.

Figure 3.41 shows the revitalization of a church roof. Note that the attic space is completely open, allowing rapid fire spread due to the absence of fire stops.

Figure 3.41 Once the roof is completed, the hidden or concealed space will provide an excellent path for fire spread.

WALL CONSTRUCTION

The fundamental structural support of a building can be provided in one of two ways. Either a structural frame is provided which supports the applied live and dead loads, or the walls are used to support the loads. Each method is widely used.

The practicality of either method is usually determined by the height of the structure. Generally, tall buildings make use of a structural frame. Economics and engineering considerations usually limit modern load-supporting walls to use in structures one or two stories in height. However, older masonry buildings have had structural supporting walls up to 70 feet (21 m) in height.

When a wall is used to provide structural support, it is termed "load bearing." Masonry buildings, either ordinary or mill, utilize load-bearing walls as the primary structural support. The interior walls may or may not be load bearing. If the interior walls serve only as partitions, then the interior structural loads will be supported by columns.

When a structural frame is used for the primary support, the exterior walls provide only the building's enclosure and are termed curtain walls or panel walls. The structural frame can be made of steel, concrete, or wood. The materials used for panel walls include glass, metal, tile, concrete block, brick veneer, wood, and various sidings.

Walls are obviously a fundamental component of a building and, similar to floors, affect the fire safety of a building in a very basic manner. Combustible walls contribute fuel to the fire. Hollow walls can provide a path for the communication of fire. Fire-resistive walls can block the spread of fire either internally or externally.

Exterior Walls

The simplest exterior wall to consider is the masonry load-bearing wall. Masonry walls can be concrete block, stone, brick, or combinations of these. Brick exterior walls perform well when exposed to fire (Figure 3.42). Masonry walls of various types provide a fire resistance of two to four hours. Although collapse of masonry walls is possible during a fire, they are typically the last structural component to fail in a wood-joisted masonry building.

Figure 3.42 Although completely gutted by fire, major portions of the exterior masonry bearing walls remain. *Courtesy of Edward Prendergast.*

If a masonry wall is otherwise structurally sound, that is, it has not been undermined or weakened, collapse during a fire will usually occur as a result of forces exerted against the wall by collapsing interior components. This will usually take place only as a result of heavy fire involvement.

Fire-resistive exterior walls do more than provide structural stability; they also tend to reduce communication of fire from structure to structure. Building codes usually require less clearance between buildings with masonry or fire-resistive exterior walls than between buildings with combustible exteriors. Of course, the existence of large window openings can still present an exposure problem.

The thickness of load-bearing masonry walls can vary from 6 inches to 20 inches (152 mm to 508 mm) (Figure 3.43). Individual bricks have compressive strengths of 1,500 psi to 8,000 psi (10 343 kPa to 55 160 kPa). However, limitations arising from different mortars, workmanship, and reinforcement reduce the maximum allowable compressive stresses from 85 psi (586 kPa) for hollow masonry to 4,600 psi (31 717kPa) for engineered masonry. The masonry in the lower portions of a wall must support the weight of the bricks above. Therefore, masonry walls in multistory buildings increase in thickness at their base.

Figure 3.43 A masonry load-bearing wall can be a combination of concrete block and brick. Note wire tie rods between the brick and block course. *Courtesy of Edward Prendergast.*

Concrete block has several classifications, including load bearing and nonload bearing, hollow and solid. A concrete block that has a cross sectional area, 75 percent of which is solid material, is classified as solid block. Aggregate used in concrete blocks includes gravel, crushed stone, air cooled slag, cinders, and shale. The fire resistance of concrete blocks varies with the thickness of the block and the aggregate used in the concrete. The best fire resistance is obtained from concrete blocks made with pumice, which is a light volcanic rock.

In addition to masonry walls constructed of bricks and blocks, it is possible to have exterior walls of precast concrete.

These walls are constructed of concrete slabs 5 inches to 10 inches (127 mm to 245 mm) thick which are usually cast at the building site.

The slabs are cast in a horizontal position and, after hardening, are lifted into position with a crane. Sometimes the slabs are positioned by being rotated about their bottom edge into the vertical plane, giving rise to the term "tilt-up construction." Such precast wall construction can be either load bearing or nonload bearing. The individual slabs can be either solid or "sandwich" panels. Sandwich panels have a layer of polystyrene one inch or two inches (25 mm or 51 mm) thick between two layers of concrete to provide better thermal insulation.

Tilt-up concrete construction provides good fire resistance. However, since a single panel can weigh 20 tons (18 metric tons), a collapse under fire conditions, such as in a building under construction, will pose a severe threat to firefighters.

Perhaps the most common type of construction is wood frame in which the structural support is provided by a framework of wooden members. The frame is made up of 2x4 or 2x6 inch (51 mm x 102 mm or 51 mm x 152 mm) vertical studs placed 16 inches to 24 inches (406 mm to 610 mm) on center. All interior structural components, such as floor joists, partitions, walls, and roof rafters are of wood.

Since it is basically an entirely combustible structure, the wood frame building poses the greatest problems to the firefighter in both interior fire fighting and potential communication to other structures. The combustible concealed spaces which unavoidably exist in the walls, floors, and attics of wood frame buildings, provide a means of fire spread once a fire enters these spaces. When a wood frame building becomes heavily involved in fire, the radiant heat generated by its combustible walls can ignite adjacent structures. To reduce these hazards, building codes require firestopping within walls and floors and impose restrictions on the allowable height and area of wood frame buildings.

Wood frame structures are of two general types: balloon frame and platform frame. In balloon frame construction the exterior wall studs are continuous from the foundation to the roof (Figure 3.44). The joists for the first floor are nailed directly to the studs without a header. In platform (or western) frame construction, the exterior wall vertical studs are not continuous (Figure 3.45). The first floor is constructed as a platform upon which the exterior vertical studs are erected. After the first-story studs are erected and braced, double 2x4's (51 mm x 102 mm) known as plates are laid horizontally along the top of the studs. The second-story framing is erected on the "platform" formed by the second-story joists and flooring.

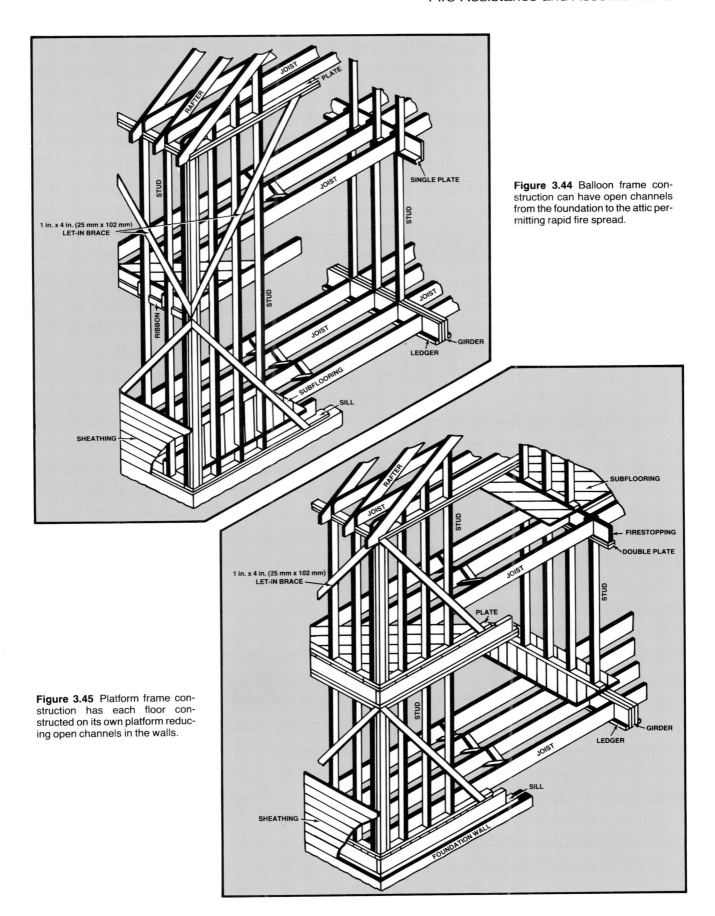

Figure 3.44 Balloon frame construction can have open channels from the foundation to the attic permitting rapid fire spread.

Figure 3.45 Platform frame construction has each floor constructed on its own platform reducing open channels in the walls.

The vertical combustible spaces between the studs in balloon frame construction provide a channel for the rapid communication of fire from floor to floor. The fire can readily spread from between the horizontal floor joists into the vertical wall and from the wall into joists on other floors. Therefore, a fire in a balloon frame building is generally more difficult to control than in a platform frame building. In a platform frame, the plate installed on the top of the studs provides a fire stop that tends to block the spread of fire from floor to floor within the walls. In balloon frame buildings, the firestopping must be provided in addition to the structural members.

From a construction standpoint, platform frame buildings are easier to erect. The flooring of each story can be used as a platform on which to work while erecting additional walls and partitions. However, greater shrinkage occurs in a platform frame than in a balloon frame. This is because the shrinkage of wood is relatively greater in a direction perpendicular to the wood grain. A platform frame building makes use of more horizontal members in the frame. This results in greater vertical movement at different points. This vertical movement causes undesirable effects, such as cracking of plaster and misalignment of door and window openings.

In addition to the structural framing, the exterior walls of a wood frame building include sheathing, siding, building paper, and insulation. The interior surface of the wall is almost always plasterboard, although in older buildings it may be plaster on lath (Figure 3.46).

Sheathing is installed to provide greater structural stability and insulation and to provide an underlayer for the siding. The most common sheathing is plywood. Siding material can include wood boards, wood shingles, asphalt shingles, asbestos cement shingles, and aluminum. Siding is available in a wide range of architectural styles. The building paper placed between the siding and the sheathing is a heavy-duty paper treated to minimize air movement through the wall.

Since the energy crisis in the early 70s, energy conservation has assumed a new importance. Increased interest has been placed in insulating building walls to reduce heat loss to the outside. Various materials are available for use as insulation in exterior wood walls. These include mineral wool, corkboard, fiber glass, shredded paper, wood fiber, polyurethane, and polystryrene. Insulating materials are available in several forms, including rigid boards, batts, blankets, and loose fill.

A well-insulated building will tend to accelerate the rate of fire growth and decrease the time to flashover. This is because a well-insulated compartment will retain more of the thermal

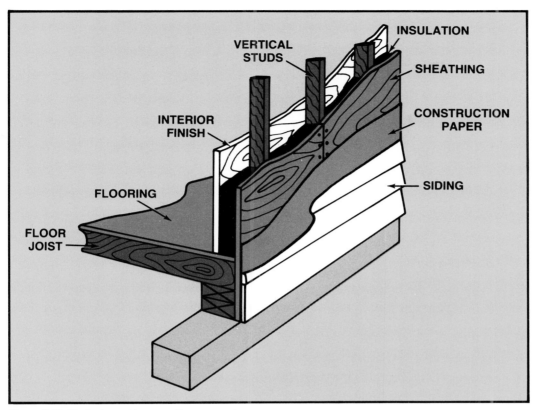

Figure 3.46 Typical wood frame wall components.

energy released by a fire within it and the temperature will rise more rapidly. See discussion in chapter 1.

The use of foam plastics as insulation has attracted considerable attention in recent years. Because they are combustible and because of the rapidity with which flame spreads over their surface, building codes impose stringent regulations on the use of foam insulation. Typically, a code will require that the foam insulation be faced with a thermal barrier, such as gypsum wallboard, to prevent or retard surface ignition of the foam.

The extent to which the presence of a foam insulation in a wood frame wall will increase fire spread within the wall will depend on the existence of an air space. If an air space exists between the foam and the wall surface, fire development within the wall will be rapid, since the fire will spread over the plastic surface and have air available from within the space. If, however, the space is completely filled with the foam, the fire would have to burn upwards through the material and will progress much more slowly. This is especially true if it is sandwiched between noncombustible coverings.[1]

---

[1]For specific technical information see NFPA 255 "Method of Test of Surface Burning Characteristics of Building Materials" and U.L.C. S124, "Standard Method of Test for the Evaluation of Protective Coverings for Foamed Plastics," Underwriters Laboratories of Canada.

The use of a combustible insulation in walls also increases the possibility of a fire starting within the wall. This is due to the possibility of an electrical malfunction igniting the insulation.

Loose-fill insulation can be made of such materials as granulated rock wool, granulated cork, mineral wool, and glass wool. The loose insulation can be either blown into stud spaces or packed by hand. Cellulose fiber and shredded wood can also be used as loose insulation material. They can be treated with water soluble salts to reduce their combustibility. However, a fire in such materials will progress in a slow smoldering manner. Thus, whenever a fire may have entered a wall space, good fire fighting tactics require that the wall be opened and the insulating material thoroughly checked.[2]

A wood frame building can be provided with an exterior facing of brick. Such construction is termed "brick veneer." Brick veneer construction provides the architectural styling of brick without the cost. The brick veneer adds little to the structural support and must be tied to the wood frame wall at intervals of 16 inches (406 mm). However, the brick veneer does add to the thermal insulating value of the wall (Figure 3.47).

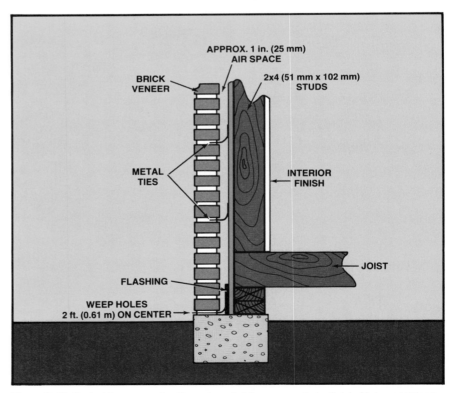

Figure 3.47 Typical frame construction using a brick veneer exterior finish. Note metal ties to hold brick to frame wall.

---

[2]The flammability characteristics of a large number of insulating materials are published in the "Building Materials Directory" published by Underwriters Laboratories.

The external brick layer protects a frame structure from external exposure. However, since the main structural support is still provided by an internal wood frame, there is little difference between a brick veneer building and an ordinary wood frame building with regard to an internal fire (Figures 3.48 and 3.49).

Figure 3.48 A brick veneer being added to a frame exterior wall. *Courtesy of Edward Prendergast.*

Figure 3.49 This platform frame apartment building awaits the brick veneer finish to provide external protection. *Courtesy of Edward Prendergast.*

From the outside it will be difficult for a firefighter to determine visually if a building has brick bearing walls or brick veneer walls. One general rule is that a brick wall will have a "header course" every sixth layer of brick. A header course is a layer of brick placed with the ends of the bricks, rather than the sides, facing out. This is done to strengthen the bond between bricks within the wall. Since a brick veneer wall has only a facing of one vertical layer, all bricks are laid with their sides exposed in what are called stretcher courses. However, some brick veneer walls will have a header course made from half bricks to architecturally simulate a regular brick wall. Therefore, no hard and fast rule can be stated with regard to identifying brick veneer walls externally.

When the exterior wall of a multistory building is a curtain wall, a potential for vertical communication of fire exists. Strictly speaking, a curtain wall is any exterior wall that is not load bearing (or supports only its own weight) and therefore can include some masonry exterior walls such as brick and clay tile. However, the term is most commonly applied to various prefabricated glass and metal assemblies (Figure 3.50). These materials provide almost no fire resistance. In some designs, the curtain wall will be entirely or almost entirely of glass, extending from the floor to the ceiling of a given story. If a fire develops on one floor of a building, the heat will crack and break the glass. If there is a sufficient volume of fire within the room, the flame plume will project through the broken window and overlap the ceiling slab above. The flame can then expose the windows of the story above, causing them to break, and allow the fire to enter the upper floor (Figure 3.51).

When a curtain wall consists of these metal and glass panels, the method of attachment of the assembly to the structural frame is of importance to firefighters. It is possible for a small open space to exist between the curtain wall and the floor slab that would permit vertical extension of fire up the inside of the curtain wall. To prevent this, suitable firestopping must be provided to maintain the continuity of the floor as a fire-resistive barrier.

Figure 3.50 A glass and aluminum curtain wall is being installed in this steel-framed building. *Courtesy of Edward Prendergast.*

Figure 3.51 Fire can spread vertically up the side of a building with glass curtain walls. *Courtesy of Edward Prendergast.*

Interior Walls

Interior walls can be load bearing or nonload bearing. They can also be fire resistive or nonfire resistive. The fire-resistive requirements and, hence, the design of an interior wall, will usually be determined by the local building code. For example, the partition wall separating adjacent apartments in an apartment building may be required to have a one-hour fire resistance. This will serve to protect the occupants of one apartment from a fire in a neighboring apartment. A common way of accomplishing this in a fire-resistive structure is to use ⅝-inch (16 mm) fire-rated gypsum wallboard applied to both sides of 2½-inch (64 mm) steel studs.[3] However, such an assembly could not be used in a load-bearing situation. If the ⅝-inch (16 mm) gypsum wallboard were applied to *both* sides of 2x4 inch (51 mm x 102 mm) wood studs, then the partition would have a one-hour fire resistance and could be used in a load-bearing application in a *wood joisted* building. It could not be used in a fire-resistive building since the structural components in fire-resistive buildings are themselves required to be fire resistive.

Building codes typically require fire-resistive partitions in locations such as corridor walls (one hour), stairway and elevator shaft enclosures (two hours), and occupancy separations (one to four hours).

---

[3]U.L. design No. U405.

When the floor area of a building is subdivided with numerous fire-resistive partitions and/or walls, it is said to be compartmentalized. The fire-resistive partitions will tend to contain fire and block its spread. Of course any openings in the partition such as doors, windows, and access panels nullify the value of the partitions unless these openings are protected by fire doors, shutters, or other means. Fire-resistive compartmentalization, while beneficial, provides only passive fire protection. That is, it may block or retard a fire but it cannot extinguish the fire. For extinguishment, either hose streams or automatic sprinklers are necessary.

Fire-resistive partitions can be constructed from a wide variety of materials, including wire lath and plaster, gypsum wallboard, concrete block, and combinations of materials. The degree of fire resistance provided will depend on the material used and its thickness. Thus, if in the previous example, two layers of ⅝-inch (16 mm) gypsum wallboard were applied to both sides of the steel studs, the fire resistance would be increased to two hours.

One useful point that can be kept in mind by the firefighter is that plasterboard partitions, while capable of providing good fire resistance, are relatively easy to penetrate with forcible entry tools. This can be helpful in reaching the seat of a fire in a compartmented building such as a modern apartment building (Figure 3.52).

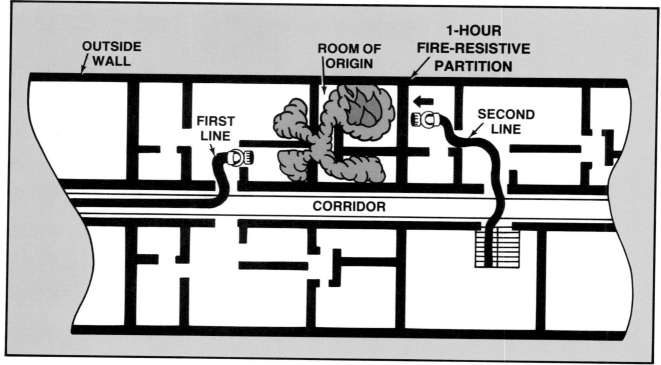

Figure 3.52 In buildings constructed along a central corridor, it is frequently possible to reach the seat of the fire by penetrating a fire-resistant plasterboard partition with forcible entry tools.

Fire Walls

The difference between a fire-resistive partition wall and a fire wall is a matter of degree. A fire wall is designed to withstand a severe fire exposure and to act as an absolute barrier against the spread of fire. A fire wall usually has fire resistance of four hours, with all openings protected by automatically closing fire doors. A fire-resistive partition will have a fire resistance less than that of a fire wall.

The purpose of fire walls is to subdivide a building into areas so that a fire in one area will be limited to that area and not destroy the entire building. For example, a 100,000 square foot (9 290 m²) factory could be divided into 25,000 square foot (2 323 m²) areas by fire walls. The containment of a fire to one area greatly reduces the economic loss and can enable a stricken industry to recover more quickly. Fire walls can also separate various functions within a plant so that loss of one will not result in total loss of a facility. For example, a hazardous chemical blending operation can be separated from shipping, warehousing, and other departments of a chemical factory.

Building codes will usually limit the maximum area of various types of buildings as a means of protecting a community from a conflagration. One municipal code restricts two-story mercantile buildings of wood-joisted construction to 8,000 square feet (743 m²) unless equipped with an automatic sprinkler system. If a larger building is desired, fire walls must be provided to limit the maximum individual areas to 8,000 square feet (743 m²).

Fire walls are always masonry. They are customarily designed to be self-supporting so that structural collapse on either side of the fire wall will not damage it. Fire walls can be constructed of brick 8 inches (203 mm) thick, solid concrete block 8 inches (203 mm) thick, hollow tile 10 inches (254 mm) thick, and combinations of these materials. Since fire walls are intended to be absolute barriers against fire, no combustible construction can be permitted to penetrate the wall. Where combustible floor and roof beams abut a fire wall, they cannot pass completely through the wall. In addition, the fire wall must extend beyond combustible walls and roofs to prevent the radiant heat of flames from igniting adjacent surfaces (Figure 3.53). This is accomplished by topping the fire wall with a parapet. The parapet height above the combustible roof will be determined by the building code. A code may require parapets 18 inches to 36 inches (457 mm to 914 mm) in height.

Figure 3.53 Masonry fire walls used to separate wood frame construction must extend beyond combustible walls and roofs. *Courtesy of Edward Prendergast.*

A properly constructed and maintained fire wall is a powerful ally to the tactical fire fighting forces. When a section of building on one side of a fire wall becomes heavily involved, the fire wall is a natural line along which to establish a defense. With fire doors closed, one or two handlines can be positioned to check for any spread of fire at cracks or around door edges. This can be accomplished with a minimum of personnel, freeing other firefighters to protect exposures or attack the main body of fire.

An opening through a fire wall can also be used as a protected vantage point from which to attack the main body of fire (Figure 3.54). However, great care must be exercised in opening fire doors in the course of a fire. Should the situation become untenable, forcing the firefighters to withdraw, fire doors must be closed.

Figure 3.54 An opening in a fire wall can be used as a protected attack position. *Courtesy of Edward Prendergast.*

FLAME SPREAD RATINGS

An important consideration in overall building fire safety is the combustibility of the materials used for interior finish, that is, the speed and extent to which flame can travel over interior surfaces. Several disastrous fires over the years have tragically illustrated that a combustible interior finish can contribute greatly to loss of life. Examples include the LaSalle Hotel fire in Chicago (1946, 61 lives) and the Winecoff Hotel in Atlanta (1946, 199 lives).

NFPA Fire Protection Handbook, 15th edition, states: "Interior finish relates to a fire in four ways: (1) it affects the rate of fire buildup to a flashover condition, (2) it contributes to fire extension by flame spread over its surface, (3) it adds to the intensity of a fire by contributing additional fuel, and (4) it produces smoke and toxic gases that can contribute to life hazard and property damage. Materials that exhibit high rates of flame spread, contribute substantial quantities of fuel to a fire, or produce hazardous quantities of smoke or toxic gases would be undesirable.

The behavior of fire in the real world is highly dynamic and influenced by several thermal variables. Precise measurement of the surface burning characteristics of materials in a manner that can be utilized in practical applications is difficult. The speed of flame spread over an interior finish material will be influenced by such factors as the chemical composition of the material used as a backing and the geometry of the space in which it is installed.

The most widely used test for determining the surface burning characteristics of interior finishes is the Steiner Tunnel Test. It was developed by A.J. Steiner, an engineer at Underwriters Laboratories. This test also has the designations ASTM E84, NFPA 255, and UL 723. When materials are subjected to this test, their surface combustibility can be expressed in a number known as a "flame spread rating." The flame spread rating provides a means of determining the hazard presented by interior surface finishes.

It should be obvious that surfaces of some materials, such as wood veneers, will pose more of a fire hazard than other materials, such as plaster. Once materials have been evaluated, restrictions can be placed on the use of those which are most hazardous. To understand how this is done a brief description of the tunnel test is necessary.

The "tunnel" in the Steiner tunnel test is a horizontal furnace 25 feet (7.6 m) long (Figure 3.55). The interior of the furnace

Figure 3.55 Horizontal tunnel furnace for determining flame spread. *Courtesy of Underwriters Laboratories Inc.*

is 17½ inches (445 mm) wide and 12 inches (305 mm) high. The top of the tunnel furnace is removable (Figure 3.56). The specimen to be tested is attached to the underside of the furnace top and the assembly is lowered into place. A gas burner located at one end of the tunnel produces a gas flame that is projected against the test material. The flame is adjusted to produce approximately 5,000 Btus (5 270 Kj) per minute. The extent of flame travel along the material is observed through view ports along the side of the tunnel.

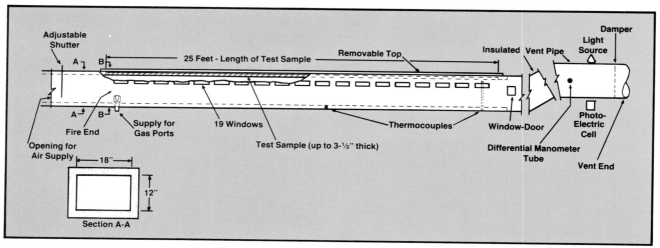

Figure 3.56 Component parts of a tunnel furnace. *Courtesy of Underwriters Laboratories Inc.*

To derive the numerical flame spread rating, the flame travel along the test material is compared to two standard materials; these are asbestos cement board and red oak. Asbestos cement board is assigned a flame spread rating of 0 and red oak is assigned a flame spread rating of 100. Obviously, the higher the flame spread rating the more hazardous the material.

The flame spread ratings of a few common materials are listed below.

Gypsum wallboard 10-15
Treated Douglas fir plywood 15-60
Mineral accoustical tile 15-25
Walnut-faced plywood 171-260
Veneered woods 515

Building codes usually classify interior finish materials according to their flame spread rating. The NFPA *Life Safety Code* uses the following classifications:

Class of Material	Flame Spread
A	0-25
B	26-75
C	76-200

A building code will specify what class of material can be used for interior finish in various occupancies or portions of a building. For example, corridors in a hospital might be limited to Class A (25) flame spread. In an office building, where the occupants are usually ambulatory, Class B (26-75) flame spread might be permitted.

In addition to the flame spread rating, the tunnel test provides two other measures of combustibility. These are the fuel contributed and the smoke developed. The measure of fuel contributed by the specimen in the test is rarely used in building codes. The smoke-developed rating is sometimes used in codes.

The smoke-developed rating of the tunnel test is not a measure of the toxicity of the products of combustion of a particular material. It is a measure of the visual obscurity created by the smoke of the given material. This is determined by passing a beam of light through the exhaust end of the tunnel furnace which impinges on a photoelectric cell.

Thus, an interior finish material might be classified on the tunnel as follows:

Material
Name
_____
Flame spread 25
Fuel contributed 50
Smoke developed 65
_____

The flame spread rating cannot be accurately determined in the field. It must be evaluated in a suitable laboratory using the tunnel furnace. However, materials can only be identified in the field, either as part of a fire prevention inspection or as part of an investigation following a fire.

Tunnel furnaces exist in several laboratories across the nation. In addition, several manufacturers of building materials have test furnaces for product development purposes. The best known source of flame spread information is the building materials directory published annually by Underwriters Laboratories, Inc.

ROOFS

The basic purpose of a roof is to protect the inside of a building from exposure to snow, wind, rain, etc. However, the roof also provides for a controllable interior environment, enhances the architectural style of a building, and contributes to the functional purpose of the building. An example of this is a domed roof over a sports arena.

From a fire fighting standpoint, there are several features of a roof which are of concern. These are

- Its ability to support firefighters
- The combustibility of its surface
- The possibility of a fire originating within the roof spaces
- The likelihood of a fire traveling through the concealed roof spaces

Roofs share some of these considerations with other building components. For example, fire travel through concealed spaces and structural integrity have been previously discussed in connection with floors and ceilings.

With respect to structural stability, it should be noted that, in general, roofs are not as strong as floors. They are usually designed to withstand a live load of 20 pounds per square foot or 25 pounds per square foot (97.8 Kg/m^2 or 122 Kg/m^2) depending on the local code. In parts of the country where winters are severe, a roof must be designed to withstand a snow load which is greater than the minimum live load. The required snow load in a given locality will depend upon the slope of the roof. As the slope of the roof increases, the required snow load decreases.

Roofs must also be designed to withstand wind forces. The local building code may require that a roof withstand a specified internal pressure to resist lifting forces that the wind may exert. When the slope of roof exceeds 30 degrees, the code will also require that the roof be designed to withstand an external wind load (pounds per square foot [kg/m^2]). The wind load will depend on the geographic location and height of the building.

Fire fighting tactics frequently require that firefighters get on the roof of a burning building. The most common reason is to ventilate the products of combustion from within the building. If a fire has not advanced to a point where it has weakened the structural support of the roof, the roof will provide adequate support for firefighters. Although roofs have to be constructed with enough strength to provide access for periodic repairs, some lightweight roofs of corrugated metal may not provide adequate support for firefighters.

When firefighters are working on the roof of an involved building, the integrity of the roof is critical to their safety. However, the extent of structural weakening that the fire has caused is very difficult to determine (Figures 3.57-3.60). Therefore, extreme caution must be used. Any indications of structural weakening, such as sagging, cracking noises, extensive fire involvement, or a fire that has burned for a prolonged period, should cause firefighters to immediately withdraw from the roof.

Figure 3.57 Firefighters at work ventilating a roof. *Courtesy of Edward Prendergast.*

Figure 3.58 After firefighters withdraw, the fire breaks through the roof. *Courtesy of Edward Prendergast.*

Figure 3.59 Fire spreads to roof structural members. *Courtesy of Edward Prendergast.*

Figure 3.60 Roof becomes totally involved and collapses. *Courtesy of Edward Prendergast.*

Roof Coverings

The combustibility of the surface of a roof is a basic concern to the fire safety of an entire community. Roofs which can be easily ignited by flaming brands have been a frequent cause of major conflagrations. The danger of easily ignited roof coverings was recognized hundreds of years ago and some of the first fire regulations in America imposed restrictions on combustible roof materials.

Test methods have been developed to evaluate the fire hazards of roof coverings. NFPA Standard 256, *Methods of Fire Tests of Roof Coverings,* describes the appropriate procedures. The test evaluates the flammability of the roof covering, the protection it provides to a combustible roof deck, and the potential for producing flaming brands. Roof materials are classified as Class A, Class B, and Class C. Class A materials possess the best fire retardant properties, Class C the least. For example, asphalt-asbestos felt-assembled sheets of four-ply thickness have Class A rating, but the same material of a single thickness would have a Class C rating.

There is a wide variety of roof coverings in use. Roof covering materials include sheet metal (galvanized steel, copper, aluminum, and lead), clay tile, slate, plastic coatings, and felt and asphalt in shingle and roll forms (Figure 3.61). Roofing is fre-

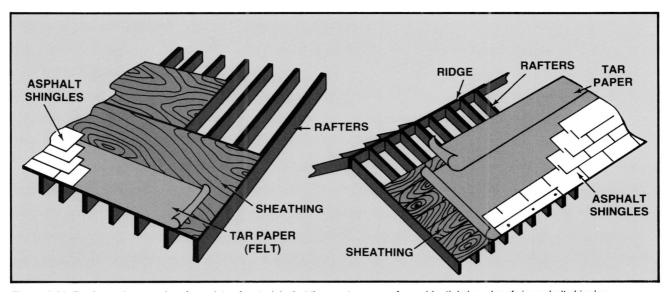

Figure 3.61 Roof coverings can be of a variety of materials, but the most common for residential sloped roofs is asphalt shingles.

quently of a built-up form. This consists of layers of roofing felt applied to a roof deck with intervening layers of roofing cement. The layers of felt will be topped with a layer of tar and gravel. Wood shingles, which are used in some parts of the country, can be readily ignited by sparks and pose a serious fire threat. Most building codes will restrict roof coverings to Class A or Class B, with Class C materials being permitted on wood frame buildings.

A combustible concealed space will sometimes exist between a ceiling and the rafters and sheathing (Figure 3.62). Sources of ignition will exist or pass through this space. Examples include chimneys, cooking vents, and electrical wiring. Should any of these be improperly installed or poorly maintained, they can become a cause of fire within the concealed space.

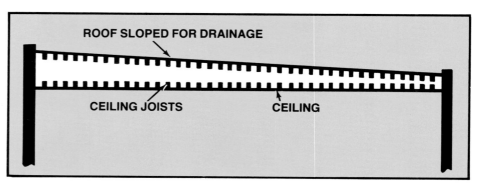

Figure 3.62 Concealed space or cockloft between ceiling and roof is often the area through which a fire spreads unchecked.

Whether a fire originates within or spreads into a roof space, it poses a difficult tactical fire problem similar to combustible floor and ceiling spaces. Usually access to the concealed space is difficult or impossible. The extent of fire spread is difficult to determine from below. It may require pulling down the ceiling well in advance of observable fire. Extreme caution is needed since the structural deterioration of the roof framing by the fire can result in collapse, trapping firefighters below.

A roof is not only a covering, it is a *waterproof* covering. This basic fact tends to frustrate efforts to control a fire in a roof space from above. The very nature of the roof tends to limit the penetration by streams from ladder pipes and platform apparatus into the seat of the fire. Should the roof collapse into the structure, it will form a waterproof covering over buried fire. This situation results in extensive damage to the structure and contents and laborious overhauling by the firefighters.

Roofs are frequently described by their style or shape. Typical roof shapes are shown in (Figure 3.63). They include the following:

Flat Roof — used on all kinds of buildings. Usually not absolutely flat but pitched slightly to permit drainage. The shape will vary from ⅛-inch to 2 inches per 12 inches (3 mm to 51 mm per 305 mm) of horizontal length.

Shed Roof — a roof which slopes in only one direction.

Pitched Roof — pitched in two directions from a high point or ridge. The slope may vary from 4 inches to 20 inches (101 mm to 508 mm) per horizontal foot (100 mm per .3 m).

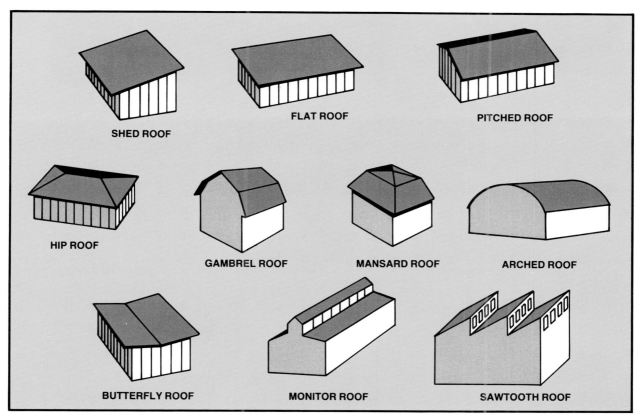

Figure 3.63 Various types of roof styles.

Butterfly Roof — two shed roofs which meet at their low point.

Hip Roof — slopes in four directions.

Gambrel Roof — slopes in two directions but has a change in the slope on both sides. This is a design that permits efficient use of the attic space under the roof.

Mansard Roof — similar to a gambrel roof with double slope on all four sides. Sometimes the central portion is flat, in which case, it is called a deck roof.

Sawtooth Roof — used in industrial buildings to facilitate ventilation and natural lighting. It consists of alternating inclined planes of different angles similar to the shape of saw teeth.

Monitor Roof — also provides for ventilation, especially in industrial buildings.

Arched Roof — a curve shaped roof constructed with masonry vault; bowstring trusses; or steel, concrete, or laminated wood arches.

Roof Supports

The structural support for a roof can be provided by a number of different methods, including joists, trusses, rigid frames, arches, domes, and cables.

Flat roofs are usually supported by roof joists. The technique is structurally similar to that used for floors. Roof joists can be wood beams or steel bars (bar joists) (Figure 3.64). The roof decking or sheathing is attached to the joists. In addition to wood, roof decks can be made of precast gypsum, concrete planks, wood-fiber cement planks, and light gage steel-ribbed panels.

Figure 3.64 Steel bar joist in place to support the roof. *Courtesy of Edward Prendergast.*

Trusses are commonly used to support roofs. A truss is a framed structural unit made of a group of triangles in one plane. If loads are applied at the points of intersection of the truss members, only compressive or tensile (nonbending) forces will result in the members (Figures 3.65 and 3.66).

Figure 3.65 Prefabricated wooden roof truss being hoisted into position. *Courtesy of Edward Prendergast.*

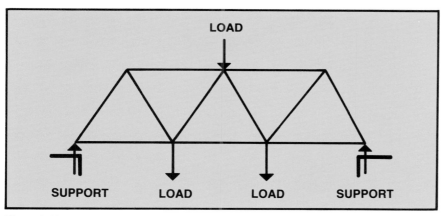

Figure 3.66 Forces in a truss under load.

A true truss is made up of only straight members. However, some types of roof trusses, such as the bowstring, have a curved top chord. These curved members must unavoidably be subjected to bending forces. In addition, loads will be applied to the truss between the intersection points of members. This also results in bending forces in the individual members.

In a truss, the top members are called the top chords (Figure 3.67). The bottom members are called the bottom chords. The diagonal members are called diagonals or web members. The joints may be formed by pin connections, welding, gusset plates, or strap connectors. Trusses may be made of wood, steel, or a combination of wood and steel rods.

The basic triangles of which trusses are composed can be arranged in a large variety of styles. The more common trusses have names and are illustrated in Figure 3.68. Standard truss shapes are available to span distances of 22 feet to 70 feet (6.7 m to 21.3 m).

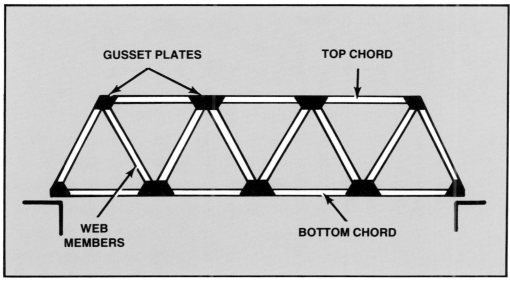

Figure 3.67 Truss components.

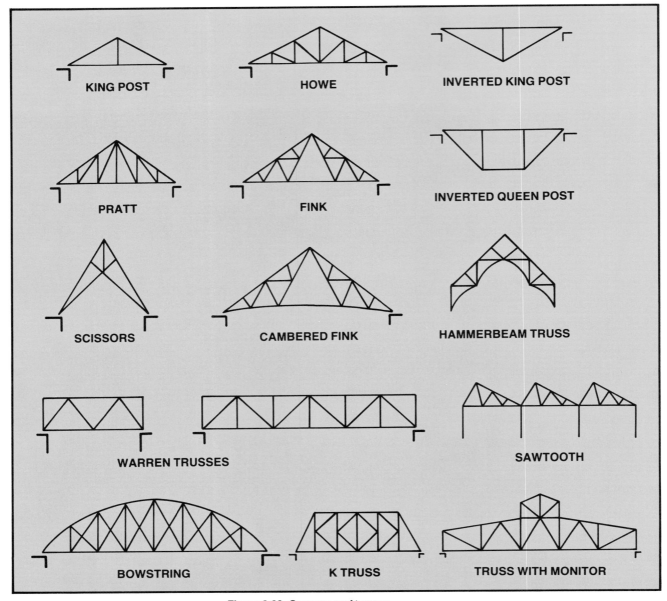

Figure 3.68 Common roof trusses.

ARCHES

Arches are structural components formed from circular segments such that the stresses induced within the arch are primarily compressive (Figure 3.69). Horizontal as well as vertical forces exist at the base of the arch. The horizontal forces must be supported by abutments or tension cables between the ends of the arch (Figure 3.70). Arches may contain hinges to permit rotation and thermal expansion. These can be provided at the top of the arch or at the abutments.

Arches are suitable for supporting roofs with large clear spaces, such as exhibition halls and field houses (Figure 3.71). Arches were originally constructed of masonry, but modern arches can be made of wood, concrete, or steel (Figure 3.72). When

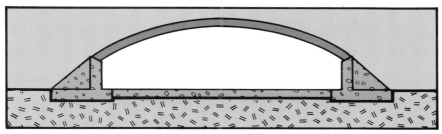

Figure 3.69 Arch roof with abutments at the end for horizontal forces.

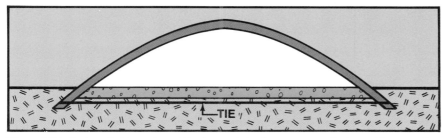

Figure 3.70 Arch roof with tension cables beneath the floor to support horizontal forces at the base.

Figure 3.71 This sports arena uses abutments to provide horizontal support to the arch roof. *Courtesy of Edward Prendergast.*

Figure 3.72 Arches are constructed of various materials. This laminated wooden arch will support a church roof. *Courtesy of Edward Prendergast.*

Figure 3.73 Laminated rigid frames are becoming more popular for roof support. *Courtesy of Edward Prendergast.*

made of steel, they can be made from plate girders or trusses. Wood arches are laminated and glued in a factory.

RIGID FRAMES

An arch has the disadvantage of restricting vertical clearance where its ends meet the floor of the building. This can be avoided by supporting the arch on a wall (Figure 3.73). However, this technique requires supports for the horizontal thrusts.

Rigid frames can provide adequate ceiling height over the entire floor area. However, they are subject to bending and shear stress. Rigid frames are sometimes called arches and do share some features with arches. Like arches, they require support for horizontal forces at their base. Also, joints may be provided at the top and base of the frame that offer little resistance to rotation. The result is a type of hinge action similar to that used in arches (Figure 3.74).

11-88

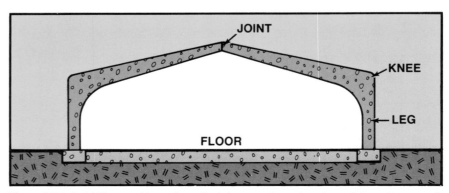

Figure 3.74 A rigid frame with ample vertical clearance has a long leg, reinforced knee, and joint at the top.

DOMES

When the area to be enclosed by the roof is circular, a dome roof can be used (Figure 3.75). A dome roof produces structural forces similar to those of an arch. That is, horizontal thrusts exist at the base and a compressive force exists at the top. However, the forces in a dome are exerted around a complete circle instead of just in one plane.

To support the outward thrust at the base of the dome and keep it from spreading, a structural member known as a tension ring is provided. At the top of the dome the forces are inward and a compression ring is used.

Similar to arches and rigid frames, dome roofs can be made of concrete, steel, or laminated wood.

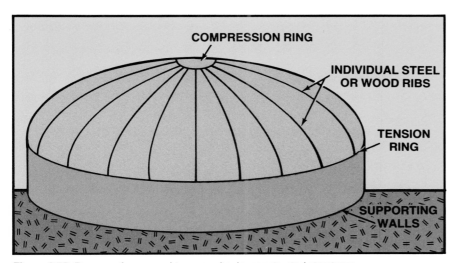

Figure 3.75 Dome roofs are used to cover circular constructed structures.

REFERENCES

"Architectural Graphic Standards," 6th edition. John Wiley and Sons.

"Building Construction, Materials and Types of Construction," 5th edition. Whitney Clark Huntington and Robert E. Mickadeit. John Wiley and Sons.

"Civil Engineering Handbook," 4th edition. McGraw-Hill.

"Design of Concrete Structures," 7th edition. George Winter, L.C. Urquhart, C.E. O'Rourke, Arthur H. Nilson. McGraw-Hill.

"Engineering Materials, Their Mechanical Properties and Applications." Joseph Marin. Prentice-Hall Inc.

"Fire Protection Handbook," 15th edition. National Fire Protection Association.

"Fire Resistance Directory," January, 1982. Underwriters Laboratories, Inc.

"Timber Design and Construction Handbook." McGraw Hill Book Co.

Building Services 4

Chapter 4
Building Services

ELEVATORS

Elevators provide the necessary transportation of people and freight in multistory buildings. In very tall structures they are more than a convenience. Today's high-rise structures reaching to 1,000 feet (305 m) or more would be impossible without a safe and reliable means of vertical transportation. Elevators are an essential operating system within a tall structure and are of critical importance during a fire.

Elevators provide a rapid access to the upper floors of a building, not only for occupants but for firefighters responding to all types of emergencies. When a fire company arrives at the front door of a high-rise structure, the company may still be as far as one-eighth of a mile (.20 km) from the fire. Straight up! If a building is of medium height (up to 15 floors), it may be possible for the hardy to bypass the elevators and walk up stairs. However, as a building's height increases to 20, 30, 50, 60 or more floors, the firefighters have no practical choice but to use the elevators. Therefore, the reliability and safety of the elevator is of extreme importance to the firefighter.

In addition to their functional aspects, the physical features and arrangements of elevators are also of importance. The shaftways can be a means for the vertical communication of fire and smoke. Building codes, therefore, have requirements for the fire-resistive enclosure of elevator shafts.

Types of Elevators

Basically, there are two types of elevators installed in modern buildings: electric traction and the electro-hydraulic. The first type has an elevator car and counterweight suspended from cables that pass over a traction sheave. The rotating sheave moves the car up and down the hoistway. In the second type, the car is mounted on a piston that moves upward as oil is pumped into the cylinder and lowers by the force of gravity as oil is released from the cylinder. A third type of elevator that is no longer manufactured but can still be found in older buildings, operates by cables winding and unwinding on a turning drum.

The use of a counterweight in a traction elevator reduces the amount of energy that must be used to hoist the car and thereby reduces the electric power consumption. A counterweight will be equal to the weight of the car plus 40 percent of the weight of the anticipated live load (passengers). On the average, the weight of the car and passengers and the weight of the counterweight will be close to equal (Figure 4.1).

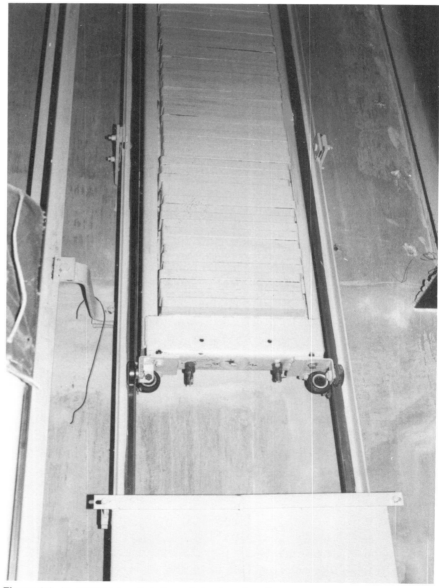

Figure 4.1 An elevator counterbalance helps to reduce the energy needed to move the car. Note guiderails on both sides. *Courtesy of Edward Prendergast.*

The traction sheave, cables, and counterweights may be arranged in a number of ways as illustrated in Figure 4.2. The sheaves indicated by S in the figure are known as the idler sheaves and can act as either a simple guide pulley or as a means of increasing the traction of the cables. The electric motor that drives the traction sheave may be located at the top of the shaft in an equipment penthouse (Figure 4.3) or in the basement. Locating the machinery in the basement requires more cable but eliminates a roof penthouse which is sometimes desirable (Figure 4.4).

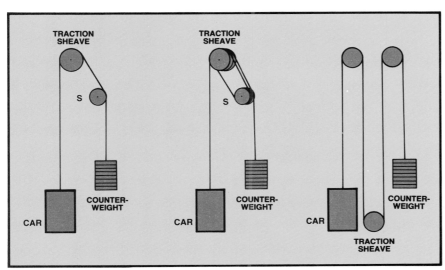

Figure 4.2 Various configurations of elevator cables, cars, and counterweights.

Figure 4.3 An elevator machine room for three elevators. *Courtesy of Edward Prendergast.*

Figure 4.4 In some cases, the machinery will be at the bottom of the shaft. *Courtesy of Edward Prendergast.*

The elevator car in a traction elevator is suspended by from four to eight steel cables. Each cable is usually capable of supporting the car by itself so that a factor of safety of seven or more can be achieved.

Elevator car speeds can be between 50 feet per minute and 700 feet per minute (15 m/min. and 213 m/min.) depending on building height. A traction elevator is usually powered by a direct current (D.C.) motor since the speed of a D.C. motor can be varied more smoothly than an alternating current motor. Since commercial electric service provides alternating current, a motor-generator set will be provided in the machine room to convert the A.C. to D.C. for each elevator. In some very modern installations, solid state control is used in place of the motor-generator to convert the A.C. to D.C.

The movement of the car is guided by rollers (Figure 4.5) on the sides of the car that move along guiderails extending up the entire height of the shaft (Figure 4.6).

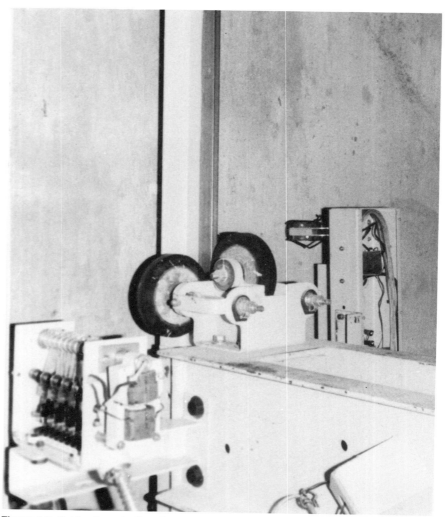

Figure 4.5 Guiderails and rollers direct the movement of the car in the shaft. *Courtesy of Edward Prendergast.*

Figure 4.6 Looking down on a car as it is hoisted, the hoisting cable is in the center and the guiderail and rollers to the right. *Courtesy of Edward Prendergast.*

Traction elevators do not have height limitations and are used in tall buildings where higher car speeds are desirable. However, where the vertical travel is modest (six stories) and high speed is not important for efficient operation, such as garden apartments and small office buildings, hydraulic elevators may be used. Hydraulic elevators are less expensive than traction elevators since they do not need cables, traction motors, and other penthouse equipment. A hole for the cylinder must be provided which is as deep as the vertical travel of the car; although, telescoping plungers exist and are sometimes used. Another disadvantage of the hydraulic elevator is that since they do not use a counterweight, the oil pump must lift the entire weight of the car and passengers resulting in higher power costs.

Elevator Safeties

Elevator development would have been impossible without the invention of elevator safeties in 1853 by Elisha Graves Otis. All traction elevators are equipped with the safety device which prevents a car from crashing into the elevator pit, even if all the hoisting cables were to break. When car speed is excessive, a speed governor will actuate, interrupting power to the traction motor and setting the operating brake (Figure 4.7). If the car continues to gain speed, the governor will activate safety jaws mounted on the car which will clamp the guiderails.

Figure 4.7 Modern direct current elevator machines have operating brakes such as the one on the right side of this machine. *Courtesy of Edward Prendergast.*

Elevator cars are equipped with travel-limited switches at the top and bottom of the shaft. If the car should travel past its normal high or low points, the limit switches cut off the power and set the operating brake.

Elevators will also have oil or spring buffers located at the bottom of the elevator shaft. These buffers are not intended to stop a free falling car. Their purpose is to cushion the stop of a car in the event that it should travel past the lowest stop. Hydraulic elevators rely on the restricted flow of hydraulic oil from the cylinder to the oil reservoir to check the fall of the car.

Hoistways

Elevator hoistways are constructed of noncombustible materials sufficient to provide a fire-resistive rating of one or two hours depending on building construction. Hoistway doors on each floor also have a fire-resistive rating to complete the enclosure of a hoistway and limit the spread of fire. Elevator hoistways in buildings constructed prior to and at the beginning of the century, however, were often not enclosed (Figure 4.8). These hoistways, if not enclosed in later years, will permit rapid fire spread from floor to floor.

Figure 4.8 Fire can spread rapidly upward in buildings with elevator hoistings that are not enclosed. *Courtesy of Edward Prendergast.*

Although modern enclosed hoistways restrict the vertical spread of fire, they fail to prevent the spread of smoke during major fires. Hoistway doors cannot fit the door opening tightly and still operate. There must be some clearance and this clearance permits smoke to filter into the hoistway. Hoistways may be vented to remove smoke, but not the quantity usually produced during major fires. This is especially true if elevator doors open on the fire floor as sometimes occur. Smoke will then enter the hoistway and filter out on the upper floors.

In buildings of modest height, the elevators generally serve all floors and the hoistway extends from the basement to the top floor. With buildings of about twenty stories and taller, elevators may be divided into zones with one zone serving the lower floors and another zone the upper floors. Hoistways of low-riser elevator banks extend from the lobby to the highest floor served by that zone, usually about midway up the building. Very tall buildings may have an intermediate-rise elevator bank, and again, the hoistway extends to the highest floor served, or about two-thirds the height of the building. Intermediate-rise and high-rise hoistways are generally "blind" from the lobby floor to the lowest floor of the zone served (i.e. there are no door openings on floors not served by the elevator) (Figure 4.9). The hoistway is enclosed on those lower floors.

A common misconception relative to elevators is that on failure of power, the cars will automatically descend to the nearest landing where exit will be possible. In actuality, the brake is set immediately on power outage and the car remains stationary. This is particularly bad for cars in blind shafts, that is, express shafts with no shaftway doors. In such cases escape from the cars via hatchway is not practicable and, when emergency power is not available, breaking through the shaftway walls is the only recourse.[1]

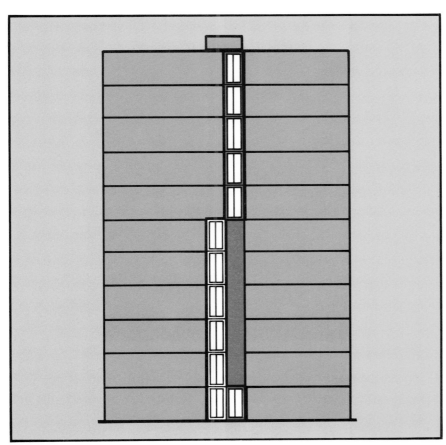

Figure 4.9 The high-rise elevator in the right hoistway serves only the upper floors and has no openings on the lower floors.

---

[1]*Mechanical and Electrical Equipment For Buildings,* 6th ed. McGuinness, William J.; Stein, Benjamin; Reynolds, John S.; John Wiley and Sons.

Elevator Doors

Hoistway doors serve two functions: they restrict the spread of fire via the elevator shaft, and they prevent people falling into the hoistway. Mechanical locks secure a hoistway door shut and the locks of all doors in a hoistway are tied together electrically (Figure 4.10). Because of this electrical tie, every door in the hoistway must be locked shut before an elevator can move. This prevents an elevator leaving a landing with a hoistway door unlocked or open. Hoistway door locks are called interlocks because of their electrical interconnection.

Figure 4.10 Elevator hoistway and car doors being installed in a two car shaft with locking mechanisms. *Courtesy of Edward Prendergast.*

Elevators are equipped with two sets of doors, one on the car and the other on the hoistway's opening. Hoistway doors are unlocked and opened by the movement of the car door. When an elevator nears a landing, the opening car door contacts a roller on the hoistway door that releases the interlock and pushes open the hoistway door. Should firefighters become stuck in an elevator between floors, they can usually escape by tripping the interlock after opening the car door. Car doors do not lock but may have to be forced open. However, should this be attempted, it must be remembered that the hoistway is open beneath the car and escaping firefighters must use extreme caution leaving the car.

The doors of freight elevators operate differently from passenger cars. The doors open vertically at the middle and are known as vertical biparting doors. This arrangement allows a full opening of the car to facilitate movement of freight. Freight shaft doors can be either manual or powered. Powered biparting doors are each powered by their own motor. Freight elevators that approach a landing contact a roller located on the side of the door opening where the door panels meet and release the interlock.

The door motor then opens the door. The two door sections move vertically with the upper panel moving upward and the lower panel moving downward. The cab itself is equipped with a gate which opens vertically. Firefighters can escape from stalled freight elevators by pushing the roller release and forcing open the biparting doors.

Some hoistway doors are equipped to be opened from the landing side of the door with a specially formed emergency key. The key, sometimes known as a "lunar" key, is inserted through an opening near the top of the hoistway door and rotated. There are several types of keys in use and the shape of the opening in the hoistway door will correspond to the shape of the key (Figure 4.11). The use of the specially shaped key is intended to prevent the opening of the hoistway door except by emergency personnel. Many modern elevator installations, however, do not have an unlocking feature except on the bottom terminal door. This terminal door provides access to the elevator pit in case of a pit fire.

Emergency exits are provided for rescue of trapped elevator passengers when rescue cannot be accomplished through elevator doors. All elevators have a top emergency hatch. Elevators in multiple hoistways may have side emergency exits, in addition to the top exit, that line up with side exits of adjacent cars (Figure 4.12). Emergency exits are designed for rescuers to get into the car and not for trapped occupants to get out of the elevator on their own. Both types of emergency exits can be opened from outside the car by rescuers without the use of a key.

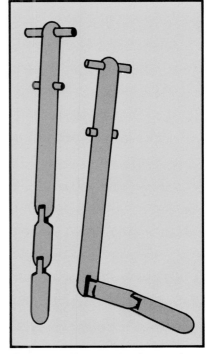

Figure 4.11 Hoistway door keys are inserted into the door and rotated to release the latch.

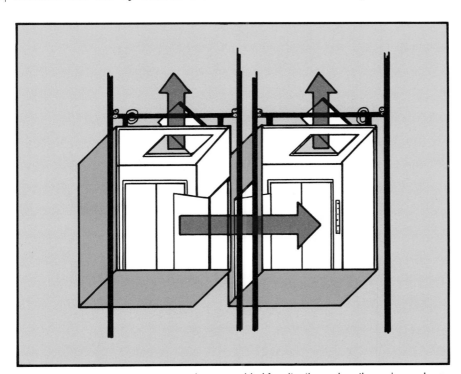

Figure 4.12 Emergency access panels are provided for situations when the main car doors cannot be operated.

Manual Control

Most elevators have a method of controlling an elevator car manually. Elevators on manual control will avoid stops at floors where hall calls have been registered. They respond only to signals originating within the car. Manually operated elevators are far safer to use during a fire than are elevators on automatic operation. Switching an elevator to manual control is usually done at the operating panel within the car (Figure 4.13). Firefighters should study their elevators to determine the correct operating procedures, as the method of operating a car manually differs among the various elevator manufacturers.

Figure 4.13 Emergency switch in the elevator car to change to manual control. *Courtesy of Edward Prendergast.*

Emergency Fire Operation

Emergency fire operation (firemen's recall) has been a requirement of the *American National Standard Safety Code For Elevators, Escalators, and Moving Walks* since 1973 and provides increased safety of elevator use during fires. The emergency operation prevents the use of elevators by building occupants during fire emergencies, the cause of fire deaths in the past, and allows greater safety for firefighters in their needed use of elevators in taller buildings.

A building code will usually require that elevators be provided with fire department controls when a building exceeds a certain height (frequently 70 feet [21 m] but in one instance ten stories). A key switch available to the fire department in the main lobby controls the operation manually, or heat or smoke sensors on each floor in the elevator lobbies activate the return of elevators automatically. When the return switch is operated, or when one of the sensing devices is activated by smoke or heat, all elevators return to the main lobby and stop with their doors open. The keys to control the elevator should be in an accessible locked box and their location indicated on the pre-fire plan (Figure 4.14).

Floor buttons in elevator cars and call buttons on each landing will no longer function, and the emergency stop button is rendered inoperative after activation of the emergency fire operation. Nothing interferes with the immediate, nonstop return of all elevators to the main lobby. This forces fleeing occupants to escape by the stairway, the safest exit from a burning building.

The key switch in the main lobby has an "on" position, an "off" position, and a "bypass" position. The "on" position activates the return of elevators, and the "bypass" position is used to shunt out the sensing device circuit when heat or smoke initiated the return of the elevators (Figure 4.15).

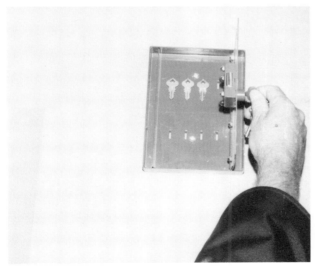

Figure 4.14 Keys for elevator emergency operation should be in a locked box, frequently in the lobby, accessible to the fire department. *Courtesy of Edward Prendergast.*

Figure 4.15 Operating the elevator emergency recall switch in the lobby adjacent to the car door returns the car. *Courtesy of Edward Prendergast.*

Elevators with a travel of 70 feet (21 m) or more will have an identical three position key switch on or beside the operating panel inside each elevator car. This car switch is operable only after the main lobby switch has been keyed to the "on" position or the return of elevators has occurred because of a heat or smoke sensing device. The car switch is used by firefighters to control the elevator manually.

Both the main lobby switch and the car switches are operated by the same key. The key can be removed from the main lobby switch with the switch "on" and then used to operate the car switch. Typical steps in the operating procedure by arriving firefighters are as follows:

- Insert key in lobby emergency switch and turn to "on" position. Cars will descend to lobby and open doors.

- Remove key from main lobby three position switch. Leave switch in the "on" position.

- Insert key into three position switch in elevator car and turn key to the "on" position.

- Register the correct floor button on the car panel. (Never press the button for the fire floor.)

- Press the "door close" button. (The button may have to be held in until the doors close fully.)

- After the car stops at the registered floor, press the "door open" button and hold the button until the door opens fully. "Door open" buttons are continuous pressure switches. They may have to be held continuously to keep the door open.

NEVER PRESS THE BUTTON FOR THE FIRE FLOOR

The continuous pressure switch for opening doors can function as a safety feature for firefighters. Should smoke enter the car as the doors begin to open on the registered floor, the doors will reclose immediately upon release of the "door open" button. Further use of the car is possible only by someone in the car. Later arriving firefighters cannot signal for the car from the lobby. The elevator responds only to signals originating from within the car.

The "bypass" position of the elevator car's three position switch is used to shunt out the safety circuits of the elevator doors. Firefighters have been trapped at a fire when they accidentally signaled the fire floor and the opening doors, warped by the heat, bound against the hoistway wall and would not reclose. Normally, elevators will not run with doors open, but the bypass permits the return of the elevator to the main lobby.

Not all elevators will have emergency fire operation. Elevators installed before 1973 do not have the emergency fire operation unless they have been modified. Also, the local governing body may not have adopted the 1973 supplement to the elevator code, in which case, the emergency fire operation will not be required. Building surveys are necessary to determine whether or not emergency fire operation is provided within a firefighter's response area.[2]

STAIRWAYS

Stairways are the basic architectural feature of buildings that provide access to and egress from levels above and below the ground floor. For stairs to serve as a safe exit for occupants in case of fire, and to restrict the vertical spread of fire, stairways inside buildings must be effectively enclosed. Building codes require that stairs be noncombustible and within fire rated enclosures, except for those in certain types of low buildings, and "monumental" stairs such as in hotels and theaters. Figure 4.16 identifies the component parts of a stairway which are basically the same as for all types of stairs.

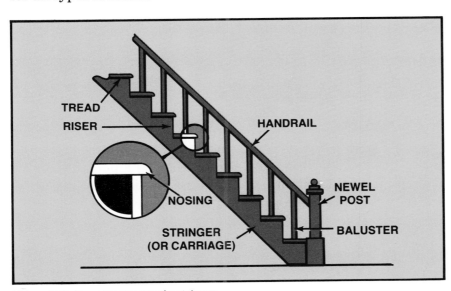

Figure 4.16 Component parts of a stairway.

Stairways can be broadly classified as interior or exterior stairs and include such stair types as straight run, return, access, scissor, folding, winding, circular, partial, moving, fire escapes, and smokeproof towers. In high-rise construction, the smokeproof tower has special significance to firefighters because it is designed to prevent entry of smoke into the stairwell from the fire floor. This enables the stairways to be used as a means of escape from floors above the fire or as a point of attack in severe fire situations.

---

[2]*Fire Department Operations With Modern Elevators,* 1977 McRae, Robert J. Brady Co., Bowie, Maryland

Straight Run Stairs

Stairs whose stringers extend in a straight line from floor to floor are called straight run stairs. They are common in small multistory dwellings and for the construction of exterior stairs in garden-type apartments (Figure 4.17). Older tenements and the old row houses usually have this type, both inside and outside. Straight run stairs usually have an intermediate platform to break up an excessively long flight of stairs. Platforms are required at intervals of 12 feet (3.7 m) vertically. Scissor stairs are usually of the straight run type.

Figure 4.17 Straight run stairs.

Return Stairs

Return stairs have an intermediate platform between floors and reverse direction at that point. Where the height between floors is greater than normal, return stairs will have more than one intermediate platform. Return stairs are very frequently used in modern construction (Figure 4.18).

Figure 4.18 Return stairs reverse after an intermediate landing.

Access Stairs

The term "access stair" refers to a function rather than a type of stair. The function is to provide access to another floor when a single tenant occupies more than one floor of a building. Employees can move more rapidly and conveniently to offices on other floors by way of access stairs than by using elevators or the regular exit stairs. This is especially true when stairwell doors are locked from the stairwell side. Access stairs are usually of the straight run type and, for asthetic purposes, are usually unenclosed. Winding stairs are sometimes used as access stairs to save space (Figure 4.19).

Figure 4.19 Access stairs, although usually straight run, can be winding.

Access stairs have both advantages and disadvantages for firefighters. The access stair is an additional opening through which to attack a fire. However, an unenclosed access stair is a path of vertical communication for a fire. It can also cause injury to firefighters who can fall down the stairway opening in a smoky fire or, as happened in a Canadian city, receive a misleading signal for a fire. In that case, smoke from a fire extended up an access stair and operated a smoke detector on the upper floor. First arriving firefighters wound up directly at the fire when they took an elevator to the floor below the signaled smoke detector.

Scissor Stairs

Scissor stairs are two sets of stairs constructed in a common stair shaft. They are cheaper to build than two separate stairways and save valuable floor space. They consist of a straight run stair which returns to every other floor. One set of stairs crisscrosses the other set of stairs and connects to the landing just opposite the first set of stairs (Figure 4.20). Egress from the stairway is into

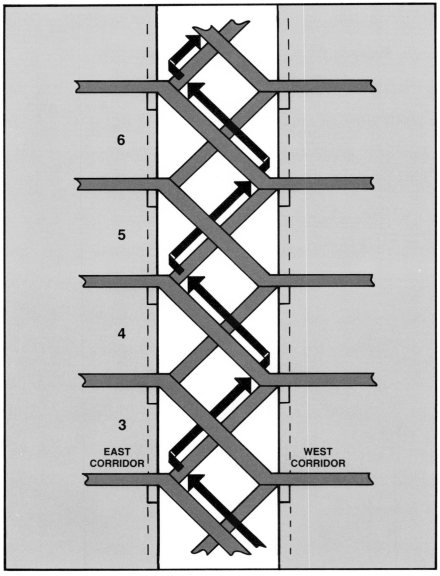

Figure 4.20 Scissor stairs crisscross to the opposite landing on successive floors.

one corridor on even numbered floors and into the opposite corridor on odd numbered floors. This can be confusing and takes additional hose to connect to a standpipe on the floor below the fire. The hoselines must cross to the other side of the center core of the building. The stairs are sealed, usually with gypsum wallboard, in such a way as to isolate the two stairways from each other. Scissor stairs are discouraged because one set of stairs is vulnerable if the other stairway becomes filled with smoke and heat. It is difficult to prevent products of combustion from penetrating one stairway after having entered the other stairway.

Folding Stairs

Folding stairs provide access to attics of dwellings when the attic is used primarily for storage, and lacks a fixed stairway. The stair is generally composed of wooden stair sections: the main section which hinges from its frame and two articulating extensions (Figure 4.21). After the stair sections are folded together, the stair

Figure 4.21 Folding stairs provide access to attic areas.

swings into the attic, and is held in place by either springs or counterbalances. It is most often located in the hallway. A light plywood panel, attached to the underside of the main section, hides the stair in its folded position. Because of this concealment, folding stairs are sometimes called disappearing stairs. Heated springs lose tension quickly and folding stairs often swing down when fire involves an attic. In this manner, it is possible for an attic fire to communicate to the living area of a dwelling. However, the folding stair provides access to the attic space and firefighters should be alert for it.

Winding Stairs

Winding stairs answer the problem of limited space. They consist of a series of steps spiraling around a single column (Figure 4.22). Each tread is tapered and connects to the column at the tread's narrower end. Steps are often suspended in cantilever style from the column. Custom-made winding stairs are usually installed in private homes, studio apartments, vacation homes, and for limited use in commercial buildings. Since they can be difficult to traverse, they are allowed for exit purposes only in limited situations.

Figure 4.22 Winding stairs require little space, but are rarely approved for exits.

Partial Stairs

Partial stairs are stairways that serve only certain floors in a multistory building rather than the entire building. These stairways may serve from 2 to 10 or 15 floors and then end. If access beyond these end points is needed, other stairs must be used. The difference between partial stairs and access stairs is that access stairs serve one tenant and usually extend only from one floor to the next. If a partial stairway does not extend to grade from the floors served, it cannot be used as an exit.

Moving Stairs

Moving stairs, more commonly called escalators, are stairways with electrically powered steps that move continuously in one direction, usually at speeds of either 90 feet per minute or 120 feet per minute (27.4 m/min. or 36.5 m/min.) (Figure 4.23). Each individual step rides a track. The steps are linked together and moved around the escalator frame by a chain known as the step chain. The driving machine is located under the upper landing (Figure 4.24) and is accessible by removing a landing plate. Con-

Figure 4.23 Escalator or moving stairs. Note sprinkler curtain wall protection around the vertical opening. *Courtesy of Edward Prendergast.*

Figure 4.24 This escalator under construction shows the machinery location under the upper floor. Note heavy steel framework. *Courtesy of Edward Prendergast.*

tinuous handrails run over the balustrades at the same speed as the moving steps. Escalators are used to accommodate large flows of people, such as crowds in office buildings, department stores, and transportation terminals (Figure 4.25). An escalator provides continuous movement of people without the delay of waiting for an elevator car to arrive.

As with any type of moving machinery associated with people, safeguards must be provided. Escalators have emergency manual stop switches or buttons located on a nearby wall out of the reach of children or near the openings where handrails disappear into the newel (Figure 4.26). Operation of the emergency switch will stop the drive and set an emergency brake. The emergency stop button can be used by firefighters when it becomes desirable to advance a hoseline up or down an escalator.

Figure 4.25 Escalators, like this one in a shopping center, move large numbers of people due to the continuous movement. *Courtesy of Edward Prendergast.*

Figure 4.26 Escalator emergency stop switch at newel base. *Courtesy of Edward Prendergast.*

Micro switches may be located behind the skirtboards (side panels of the balustrades) and under the combplates to shut off the electric power if someone gets caught in the moving steps. Improved design, however, has largely mitigated this hazard. The modern step's cleats, for example, have been narrowed and are "combed" as the steps enter a landing to dislodge soft-soled shoes. Escalators are usually not enclosed and require provisions to limit the vertical spread of fire. Automatic rolling shutters are sometimes used to close the escalator opening in case of fire. More often, closely spaced automatic sprinkler heads in the ceiling around the opening provide this protection. Accompanying curtain boards located at the ceiling and surrounding the opening aid in banking the heat to fuse the sprinkler heads quickly.

Fire Escapes

Fire escapes are a series of steel balconies and stairs mounted on an exterior wall to provide an emergency exit. Fire escapes are used on buildings lacking sufficient interior stairways (Figure 4.27). Usually, the lowest stair section is stored above ground to prevent the blockage of a sidewalk or alley, or to prevent illicit access to the building. The stair section pivots and swings down from the weight of a person stepping onto it (Figure 4.28). Some

Figure 4.27 An additional exit for emergency use is provided by a fire escape on the rear of this office building. *Courtesy of Edward Prendergast.*

Figure 4.28 A typical fire escape with a counterbalanced bottom section and a vertical ladder extending to the roof. *Courtesy of Edward Prendergast.*

fire escapes terminate above ground and escaping occupants must either jump or await arrival of the fire department to place portable ladders. A steel ladder extends from the top balcony to the roof. Balcony floors and stair treads may be open bar or solid. The latter is preferred because it enhances the sense of security of people using the fire escape.

Wall openings near fire escapes should be protected against fire, but this protection is not always provided. Flames from an exposing window or doorway can trap people coming down the fire escape. Wire glass window panes can offer sufficient protection to enable occupants to clear the fire escape before fire breaks through a window. Firefighters should be cautious in their use of fire escapes where exposing windows and other openings lack minimal protection.

Neglected fire escapes may present another hazard. Fire escapes are exposed to the elements and if not periodically painted and inspected, will deteriorate structurally over the years. Firefighters should always be alert for loose anchor bolts and rusted stair components. Especially dangerous is a deteriorated vertical ladder. Firefighters should avoid gripping the rungs when climbing the vertical ladder to the roof; they should grip the beams instead.

Smokeproof Towers

An enclosed interior stairway is a means of escape for building occupants on floors above a fire. However, if the stairwell becomes filled with products of combustion, it will be unusable, raising the spector of occupants being unable to flee the structure.

Stairway enclosures that have exterior balconies or vestibules open to the outside air or those with smoke shafts are known as smokeproof towers. An occupant must go outside the building onto the balcony or pass through a vestibule to reach the stairway. Vestibules will have at least a 16-square foot (1.5 m^2) opening through the exterior wall (Figures 4.29a and b). With either of these arrangements, smoke is prevented from entering the stairwell when corridor doors are opened. Smokeproof towers are the safest means of exit from multistory buildings.

The center core construction of modern buildings makes it costly to provide balconies or vestibules on an exterior wall and two modified versions of the smokeproof tower are frequently used. In one version, the vestibule opens to a smokeshaft that extends the height of the building and is open to the atmosphere. The opening between the vestibule and the smokeshaft is equipped with a shutter that opens automatically. Any products of combustion that enter the vestibule are drawn into the smokeshaft and the stairwell is protected (Figure 4.30). A second

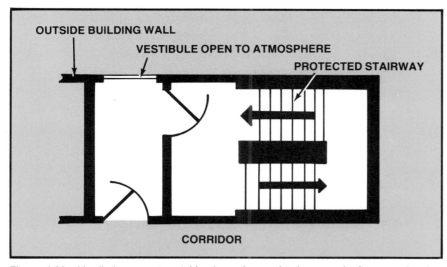

Figure 4.29a Vestibules open to outside air can be used to keep smoke from entering the stairwell.

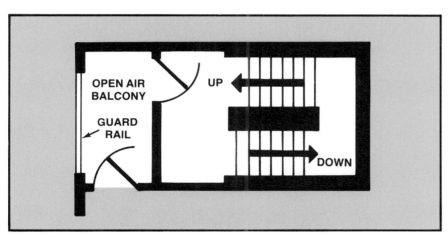

Figure 4.29b An open balcony can be designed to vent smoke away from enclosed stairwells.

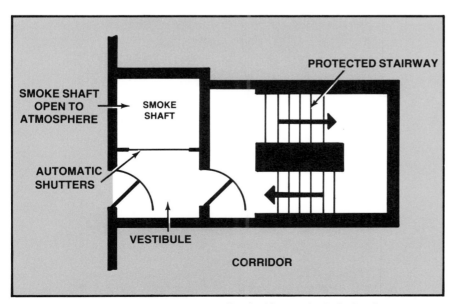

Figure 4.30 Internal stairway can be protected by providing smokeshafts that direct smoke to the atmosphere.

version incorporates a pressurized stairwell and a mechanically ventilated vestibule in the center core. Modern building codes will typically require that one of the stairways be of the smokeproof type in high-rise structures.

The pressurized stairwells of smokeproof towers have intake and exhaust fans of different cubic-feet-per-minute (m^3) ratings. The exhaust fan has a lower capacity than the intake fan. This imbalance produces a positive pressure within the stairway and prevents entry of smoke. With all doors closed in the stairway, a minimum pressure of .05 inch (1.27 mm) of water column is developed. This pressure requires a force of about 10 pounds (4.5 kg) to open a stairway door. Smoke detectors immediately outside the enclosure on each floor will activate the fans.

Although pressurized stairways can be found without accompanying vestibules, a mechanically ventilated vestibule is a desirable adjunct for smokeproof stairs. Air is supplied to and exhausted from the vestibules through separate ducts (Figure 4.31). Exhaust registers are located near the ceiling of each ves-

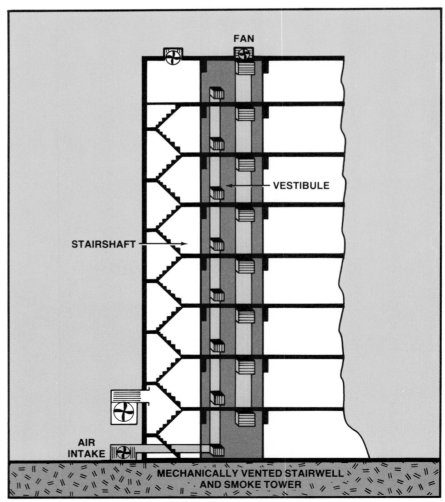

Figure 4.31 Mechanically ventilated smokeproof towers are an efficient way to protect stairwells.

tibule and intake registers near the floor. There may be only one fan on the exhaust side of the system or one on both the exhaust and supply sides. The smoke detectors that activate the stairway fans also activate the fan or fans for the vestibules.

Smokeproof stairways permit firefighters to begin fire fighting procedures before stairways above the fire floor are cleared of escaping occupants. With ordinary stair enclosures, fire fighting may be delayed until people are out of the stairway. The positive air pressure in the smokeproof stairway greatly reduces the entry of smoke when hoselines block open a stairway door. This is possible even with the simultaneous opening of three different doors.

The exhaust system of vestibules (whether mechanical or natural) can aid in venting smoke and heat from the fire floor. However, that is not its main purpose and other methods of ventilation may be necessary to facilitate fire fighting.

UTILITY CHASES AND VERTICAL SHAFTS

Utility chase is a term generally applied to the vertical pathways in a building through which pass the "lifelines" of a building. These include plumbing, electrical raceways, communications cables, and ductwork for heating, ventilation, and air conditioning (HVAC) (Figure 4.32). In addition, vertical shafts are pro-

Figure 4.32 This vertical shaft in a building under construction is part of the ventilation system. Note size of opening. *Courtesy of Edward Prendergast.*

vided for refuse chutes, linen chutes, light shafts, and material lifts. These utility and vertical shafts are important because they not only can provide a path for the vertical communication of fire and smoke but because fires can also originate within them.

In 1961 a fire that originated in the hospital linen chute in Hartford, Connecticut resulted in the loss of 16 lives. This section will address the chases, shafts, and chutes found in buildings and their fire protection characteristics.

Pipe Chase

A pipe chase is a type of utility chase which is used for the various piping needed for a building. These can include hot and cold water lines for sinks and toilets, sanitary drain lines, hot water for heating, and sprinkler risers. Depending on the size and needs of a building, one or several pipe chases can be provided.

As with any vertical opening in a building, pipe chases can spread fire and smoke to other floors of the building. Shaft enclosures should be constructed of fire-resistive materials with access openings protected to limit fire spread. Some older buildings may be found with wooden access panels covering the shaft openings that offer little in the way of protection against fire spread. A fire in the Empire State Building in the early 1960s spread 42 floors, feeding on combustible insulation of the steam pipes, and burned out of the pipe chase on two separate floors. The combustible covering of electrical wiring and communication cables pose a similar fire problem when located in vertical shafts.

Modern buildings often do not have pipe chases as such, but rather utilize the mechanical equipment room as a kind of pipe chase. Pipes and electric raceways will pass unprotected through mechanical and electrical rooms that have common locations on each floor, one above the other. The fire-resistive walls and door assemblies of the mechanical rooms are relied upon to provide the required separation of pipe chases from the remainder of the building. Additionally, holes in the floor through which pipes and raceways pass are packed with a noncombustible material to restrict the vertical spread of fire (Figure 4.33). This vertical protection usually does not exist in ordinary pipe chases. Furthermore, some buildings can be found where floor holes lack the packing or where the opening in the floor was not designed for packing.

When the ceiling space of a building is used as a return air plenum for the HVAC system, a common practice in modern buildings, fire can extend out of or into the mechanical room. Construction cost is reduced when conditioned air returns through the open ceiling space to the air handler in a mechanical room rather than through separate ducts. Fire dampers should be installed in the wall of the mechanical room above the suspended ceiling to prevent the spread of fire, but this safeguard is some-

Figure 4.33 Pipes passing through the concrete floor of the building have noncombustible fire stopping material placed around them. *Courtesy of Edward Prendergast.*

times omitted. In fact, walls of some mechanical rooms do not extend above the suspended ceiling. The mechanical room is open to the entire ceiling space of each floor.

Distribution of electric and telephone service is accomplished via the ceiling space. Electric conduit and communications cables pass through the ceiling space and to terminal points on the floor above (Figure 4.34). "Poke-throughs" provide for the passage of cables and wires through a floor. Failure to properly pack the space around a poke-through has been responsible for extension of fire from floor to floor.

Figure 4.34 Electrical cables passing from the electrical equipment room into the ceiling space. *Courtesy of Edward Prendergast.*

It is not uncommon to provide a separate chase for telephone cables to telephone closets on each floor. The possibility of fire spread through the chase and out of the telephone closet into a ceiling space can be similar to that of the mechanical room. Occasionally, louvers are installed in the entry doors of telephone closets. Such louvered doors were responsible for the spread of fire at the New York World Trade Center in 1975. The fire entered a telephone closet on the eleventh floor through a louvered door, extended through the chase, and out the louvered doors on the tenth and twelfth floors. When fire involves one of these chase-forming rooms, mechanical rooms, or telephone closets, adjacent ceiling spaces should always be inspected for fire spread.

Plumbing pipes in residential and small commercial buildings of wood frame and ordinary construction form a path through which fire can spread. Plumbing fixtures drain into a vertical pipe which connects to the underground sewer pipe and extends above the roof to ventilate the system. How tightly the stack pipe fits the hole in the top or bottom plates in platform construction, or how well the channel is fire blocked in balloon construction, determines the rapidity of fire spread through the wall space (Figure 4.35). Additionally, horizontal runs of drain pipes to bathrooms on the upper floors may be placed between the joists that form the ceiling space. This provides a lateral path from the vertical chase for the extension of fire. All bathroom areas above the ground floor must be checked when fire involves a wall containing a plumbing stack.

Often 4-inch (102 mm) pipe is used in making a plumbing stack. When 4-inch (102 mm) pipe is used, the wall must be constructed of 2 x 6 inch (51 mm by 152 mm) studs to accommodate the pipe, or as is often done today, two standard 2x4 inch (51 mm by 102 mm) studs must be used together. This forms a double wall. In double wall construction, the studs are staggered alternately creating an opening for the horizontal spread of fire.

Figure 4.35 Lateral and vertical fire spread is possible around these plumbing pipes. Note the size of the unprotected openings around the pipes. *Courtesy of Edward Prendergast.*

Refuse Chutes

Refuse chutes provide for the removal of trash and garbage from upper floors of buildings. A large vertical chute extends through the building and has openings on each floor for the deposit of trash (Figure 4.36). The chute terminates at ground level or in the basement where the trash is hauled away or burned in an incinerator.

Improperly constructed refuse chutes create a severe fire problem. The trash deposited in the chute is largely combustible, and it is easy for a lit cigarette to be dropped into the trash. Frequently, occupants will attempt to force objects into the refuse chute that are too large. This can result in a jam of material in the chute. For these various reasons, fire protection for refuse chutes is very important. Refuse chutes are required to be of noncombustible material. The opening doors are fire rated and bear the label of a fire test laboratory (Figure 4.37). The chute itself will be surrounded by a fire-resistive enclosure similar to a stairwell. Sprinkler heads will be installed at various floor levels and at the bottom where the trash is collected.

When a fire occurs in a properly protected refuse chute, it can usually be contained within the chute. However, the smoke will frequently escape onto the various floors despite operation of sprinklers and the chute doors (Figure 4.38).

Figure 4.36 This refuse chute under construction shows the first floor termination point. A fire-resistive enclosure will be provided. Note piping on the right that is the water supply for the chute sprinkler system. *Courtesy of Edward Prendergast.*

Figure 4.37 Labeled refuse chute door. *Courtesy of Edward Prendergast.*

Figure 4.38 Chute doors will contain the fire, but not stop the spread of smoke. *Courtesy of Edward Prendergast.*

Linen Chutes

Linen chutes are similar to refuse chutes in construction and potential fire spread. They provide for the removal of soiled linen from upper floors of hotels, hospitals, and similar occupancies. Requirements to safeguard against fire spread are similar to those of refuse chutes for enclosures and fire protection.

Light Shafts

Light shafts or light wells are vertical shafts through which natural lighting and air can be admitted to a building. They may be located within a building to provide light and ventilation for interior rooms or placed on exterior walls where buildings adjoin one another. With exterior walls, a portion of the wall is set back to form the well. Each adjoining room has windows that open to the light shaft. Such arrangements are usually associated with older buildings erected before the advent of air conditioning (Figure 4.39).

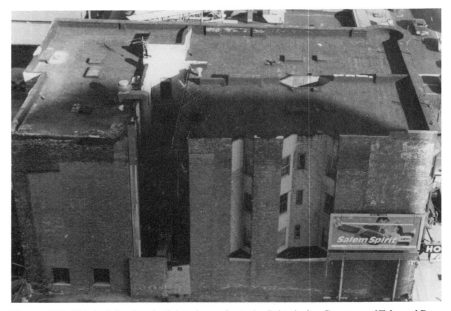

Figure 4.39 This building has both interior and exterior light shafts. *Courtesy of Edward Prendergast.*

Light shaft windows commonly lack protection against fire spread in the older buildings. Fire that has extended into a light shaft can quickly spread back into the building on upper floors. With a common light shaft such as a single shaft shared by two adjoining buildings, the spread from building to building is possible. Access to light shafts is usually not possible from the street. Because of this blind arrangement, it is impossible to observe a spreading fire in a light shaft. A check of the roof is necessary to discover a light well in a burning building. The bottom of light shafts also provides a place for the accumulation of trash which can be ignited by a dropped cigarette.

EMERGENCY ACCESSIBILITY

Before the advent of fluorescent lighting and air conditioning, buildings were constructed with large areas of windows to provide natural light and fresh air. Industrial and warehouse roofs were of the saw-tooth design or had skylights and monitors to aid in lighting the interior of buildings. There were ample openings for ventilation. Today, windows and skylights are no longer needed for the comfort of building occupants. In addition, concern for energy conservation has prompted construction with a minimum of exterior openings. The blank walls and solid concrete roofs of modern construction make it difficult to ventilate a building.

The blank walls typical of modern industrial and mercantile buildings also make access to the building difficult. Therefore, buildings are sometimes provided with openings other than normal doorways and windows through which firefighters can gain access or accomplish ventilation quickly. Windowless windows, access panels, roof hatches, smoke and heat vents, and skylights are openings that can be used to overcome the limited accessibility of buildings.

Windowless Windows

The term windowless window applies to a single layer of brick or similar material that covers an opening in a wall. It is designed so the covering can be breached easily for access into a building. The openings measure approximately 3½ feet by 3 feet (1.06 m by .9 m). The location, materials used, and methods of construction for windowless windows should be identified during prefire planning (Figure 4.40).

"STACK BOND" OF 4 in. x 8 in. x 16 in. (102 mm x 203 mm x 406 mm) CONCRETE BLOCK IN A SOLID WALL, OR SINGLE THICK BRICK.

Figure 4.40 Typical windowless window construction using a single layer of concrete blocks four inches by eight inches by sixteen inches (101 mm by 203 mm by 406 mm).

Access panels are similar to windowless windows except that the opening is covered by a removable panel rather than a fixed material which must be breached. Access panels are usually located in each story above ground level in windowless buildings. They may be required in the blank walls that face a street. According to most building codes, the openings, where required, must measure at least 32 inches (813 mm) wide and 48 inches (1 220 mm) high with sills no farther than 32 inches (813 mm) above the floor. An access panel would usually be required every 50 linear feet (15 linear meters) of wall on each floor (Figure 4.41).

Figure 4.41 Access panels for the second floor of a merchandise warehouse. *Courtesy of Edward Prendergast.*

It must be remembered, however, that any feature of a building that facilitates emergency access by the fire department can also provide a point of illegal entry by intruders. The need for building security and desirable fire protection sometimes conflict. In high crime areas or occupancies with "target" merchandise, property owners are understandably reluctant to make access to their buildings easy. Where this is a problem, firefighters must simply resort to the techniques of forcible entry.

Roof Hatches

Roof hatches provide access to a roof from inside the building and, conversely, can be used to get into a building from the roof or used for ventilation. In multistory buildings, roof hatches are usually placed above the stairway. Some type of lock, usually a padlock, secures the hatch to prevent illicit entry into the building or unauthorized access to the roof. If possible, the entire frame should be removed from its curb when a roof hatch is opened for ventilation or emergency access of a building. It is difficult and far more damaging to attempt to force the lid when a roof hatch is secured.

Stairways are used quite often to ventilate high-rise buildings that have fixed windows. The roof hatch releases smoke and heat from the stairwell and should be opened before the fire floor is vented to stairs. It is sometimes difficult to force the lock of a roof hatch when the key is not immediately available because of the cramped space at the top of the stairs. One building manager solved the unavailable key problem by using a combination lock with the combination corresponding to the street address of the building.

It must be remembered, however, that when a hatch located at the top of a stairway is used for ventilation, it will render the stairway useless as an exit from floors above the fire.

Frequently, smoke will unavoidably enter a stairwell when the stairwell doors are opened for fire fighting or escape. It may become desirable or necessary to remove the smoke through the use of ventilation fans placed at the top of the stairwell.

Smoke and Heat Vents

Smoke and heat vents, sometimes referred to as fire vents, provide for the ventilation of modern buildings in case of fire. Unit-type heat and smoke vents are small area hatchways (a minimum of four feet [1.2 m] in either direction to be effective) with single or double leaf metal lids or plastic domes designed to open automatically (Figure 4.42). There are unit vents that must be

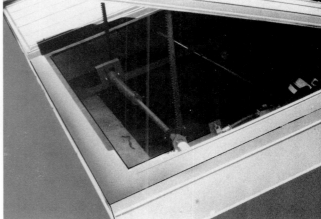

Figure 4.42 Automatic heat vent installed in roof to remove heat and toxic smoke. *Courtesy of Edward Prendergast.*

opened manually, but automatic fire vents are more desirable because they prevent delay in ventilation. Automatic action is initiated by fusible links or a tie-in with an automatic sprinkler system or heat activating devices. Counterweights or spring loaded levers cause the vents to open after being released. Because of the possibility of electric service interruption during fires, electric power is avoided as a means of opening fire vents automatically.

Monitors are single unit vents with louvered sides for ventilation or with glass sides that will break out from the heat of a fire. The glass, however, does not always break quickly enough, and firefighters will have to assist ventilation by breaking the glass manually. Some side panels are hinged and swing down when released for automatic venting.

Gable-type structures are often ventilated by continuous slot openings along the ridge of the gable. A hood protects the opening from the weather. Sometimes the throat opening of the vents will be dampered and fusible links and counterbalances provide for automatic opening.

The use of plastic panels for smoke and heat venting is relatively new. The plastic panels, which have a low melting point of 300°F to 400°F (148°C to 204°C), melt out from the heat of a fire. This automatic method of venting has proven effective when fire develops rapidly and burns intensely at high temperatures; the type of fire usually associated with severe fire hazards.

The ventilation principle in fire fighting of one large hole being better than several holes does not apply to smoke and heat vents. It is better to have a number of small fire vents uniformly distributed than to have one large vent. Chances are improved with many small vents that a vent will be near any incipient fire and faster ventilation will occur. The National Fire Protection Association publication, *A Guide For Smoke and Heat Venting* (NFPA 204), recommends that this distribution be based on the ratio of vent area to floor area in each of three basic types of occupancies (Table 4.1). This publication also includes a maximum spacing requirement for vents.

TABLE 4.1
SPACING REQUIREMENT OF VENTS

Occupancy (Contents)	Required Vent Areas	Maximum Spacing Between Vent Center to Center
Low Heat Releasing	1 sq. ft. (0.09 m²) per each 150 sq. ft. (13.9 m²) floor area	150 ft. (46 m)
Moderate Heat Releasing	1 sq. ft. (0.09 m²) per each 100 sq. ft. (9.3 m²) floor area	120 ft. (36 m)
High Heat Releasing	1 sq. ft. (0.09 m²) per each 50 sq. ft. (4.6 m²) floor area	75 ft. to 100 ft. (23 m to 30 m)
	1 sq. ft. (0.09²) per each 30 sq. ft. (2.8 m²) floor area	Depending on Fire Potential

Examples of low heat release occupancies include metal stamping plants, foundries, and breweries. Moderate heat release occupancies include automobile assembly plants, leather goods manufacturing, and printing plants. High heat release occupancies include chemical plants, rubber products plants, and general warehouses.

In large undivided areas, proper spacing of vents is not enough; the lateral spread of hot gases must be restricted by compartmenting ceiling areas in order for roof vents to operate effectively. Curtain boards form these ceiling compartments to trap heat gases so vents will open quickly. The barriers are noncombustible and extend downward from the ceiling. To be effective, curtain boards should have a minimum depth of six feet (1.8 m) for moderately high ceilings, and greater depths for higher ceilings and for curtain boards surrounding special hazards.

In 1953, General Motors transmission plant at Livonia, Michigan was destroyed after fire raced unchecked through the unvented building which totaled almost 1.5 million square feet (139 350 m²) of undivided space. Firefighters were prevented, in part, from controlling the fire because of the intense heat and smoke accumulation. The lack of effective roof venting was advanced by authorities as one of the major reasons for the complete destruction of the plant. Automatic fire vents are most suitable for such buildings (i.e., single-story, undivided buildings of large area). Windowless buildings and underground structures can also utilize smoke and heat vents advantageously.

Skylights

Skylights provide natural lighting to the interior of buildings and may be used to ventilate heat and smoke in the event of fire. Firefighters must remove a skylight or break out the glass panes to ventilate a building as skylights do not usually have provision for automatic venting. Most building codes require wired glass or tempered glass in skylights. Modern skylights use plastic domes or other shapes (Figure 4.43). Older skylights should be removed in preference to breaking the glass. The flashing of a skylight is nailed or screwed to the curb. Skylights may be removed after prying loose the flashing. Loosening the flashing on three sides will permit the skylight to be hinged over the remaining attached flashing. If skylights are completely removed, they should be left upside down beside the hole. An inverted skylight is a warning to firefighters of an opening in the roof.

Figure 4.43 A variety of new and older skylights of glass and plastic construction are in use on these building roofs. *Courtesy of Edward Prendergast.*

REFERENCES

"Architectural Graphic Standards," 6th edition. John Wiley and Sons.

"Building Construction, Materials and Types of Construction," 5th edition. Whitney Clark Huntington and Robert E. Mickadeit. John Wiley and Sons.

"Civil Engineering Handbook," 4th edition. McGraw-Hill.

"Design of Concrete Structures," 7th edition. George Winter, L.C. Urquhart, C.E. O'Rourke, Arthur H. Nilson. McGraw-Hill.

"Engineering Materials, Their Mechanical Properties and Applications." Joseph Marin. Prentice-Hall Inc.

"Fire Protection Handbook," 15th edition. National Fire Protection Association.

"Fire Resistance Directory," January, 1982. Underwriters Laboratories, Inc.

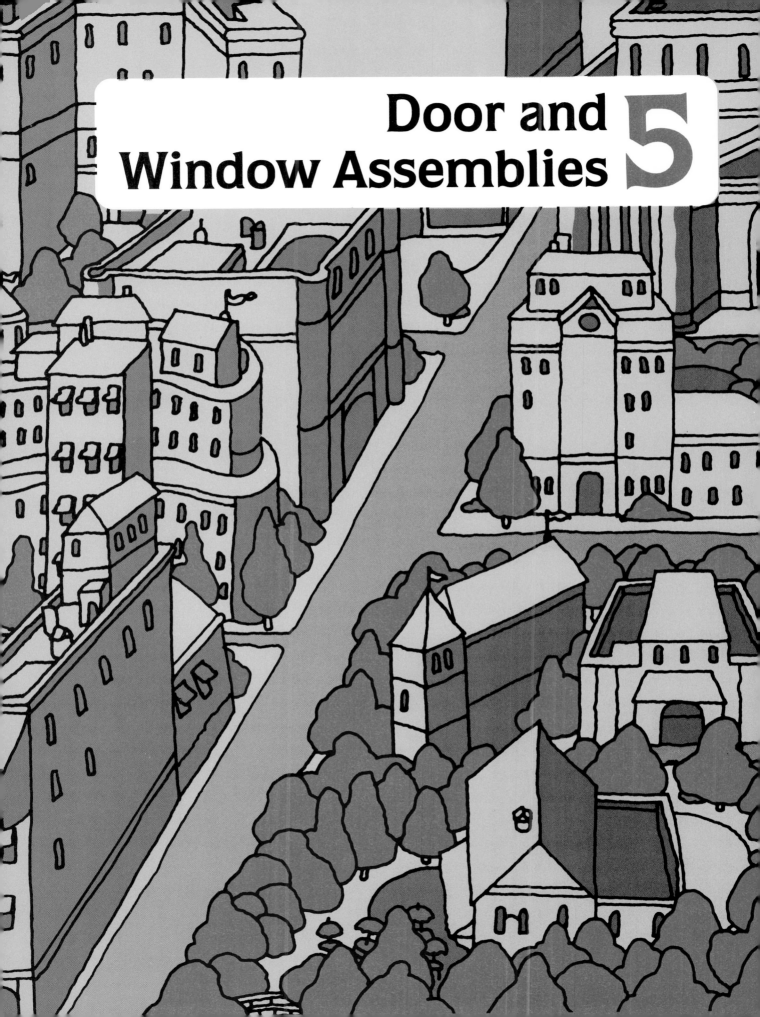

Door and Window Assemblies 5

Chapter 5
Door and Window Assemblies

DOOR ASSEMBLIES

Doors vary in design, construction, and application. The primary purpose of a door is to control access into and out of an area of separation and to prevent the weather elements from entering into a building. In addition, doors may provide a source of natural light and ventilation. However, doors can and have been a hindrance to fire fighting operations. Therefore, it is imperative that firefighters have an understanding of door construction and associated hardware.

This section will address the types of doors a firefighter may encounter. The doors addressed will be classified by their type of operation. These include swinging doors, sliding doors, folding doors, rolling doors, slab door, and revolving doors.

Swinging Doors

Swinging doors typically operate on hinges secured to the side jambs, although they may also operate on pivots supported by the head jamb and floor. The majority of doors are hinged (Figure 5.1). Pivot hardware is used mainly for double-acting doors.

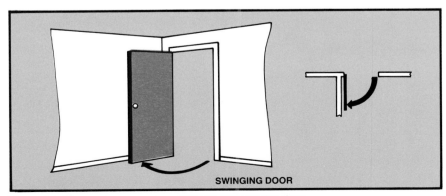

SWINGING DOOR

Figure 5.1 The most common type of door is the swinging door.

Hinged operation provides the greatest degree of security, weather resistance, and heat insulation. The operating feature of the swinging door is most convenient for entry and passage avenues which are subject to frequent use.

Door swing is referred to as being either right- or left-handed. A right-hand swinging door, whether in-swinging or out-swinging, is one that hinges from the right side of the door casing as one enters the room or building. By the same token, a left-hand swinging door is hinged on the left side.

Sliding Doors

Sliding doors are suspended on overhead tracks and operate on either alloy steel or nylon rollers. Floor guides are usually provided to prevent the door from swinging laterally. Sometimes sliding doors are mounted on rollers operating on floor tracks, but these are not common.

Sliding doors are suitable for visual screens as in closets and storage areas where bypass sliding (Figure 5.2) is common. Sur-

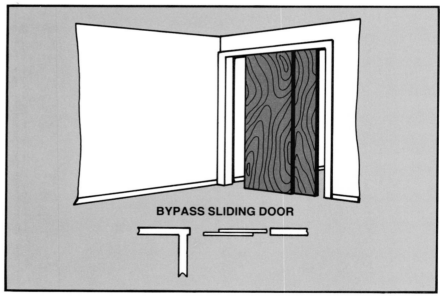

BYPASS SLIDING DOOR

Figure 5.2 Bypass sliding doors are commonly used on patios and as lightweight interior doors.

face sliding and pocket sliding doors (Figure 5.3) that operate in a similar manner may be used for passage doors. Biparting surface or pocket sliding doors are used for openings of three feet (.9 m) or greater. For extremely large openings, multi-track installations permit stacking of several sliding doors at one or both ends.

Sliding doors do not give a high degree of privacy, sound insulation, or weather resistance unless custom hardware or pre-engineered door units are used. The chief advantage of sliding operation doors is that it eliminates door swings that might interfere with the use of interior space.

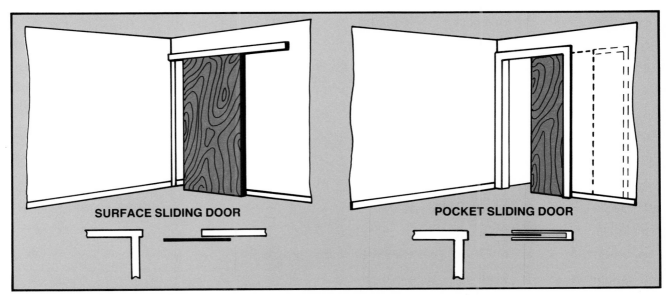

Figure 5.3 Surface and pocket sliding doors are used to eliminate the space requirement for a door to swing.

Folding Doors

Folding doors are hung on overhead tracks with nylon rollers or glides similar to sliding doors. Some types also require a floor track. Folding-type doors, like sliding, are generally appropriate for locations requiring visual screening primarily.

Folding doors are often used for closets in single or paired bifold arrangements (Figure 5.4) or as space dividers and movable partitions with special hardware for multifolding action and stacking (Figure 5.5). Accordion folding doors are used as closet doors or as space dividers and may be pocketed or end-stacked within the opening.

Figure 5.4 Bifolding doors have similar applications to sliding doors.

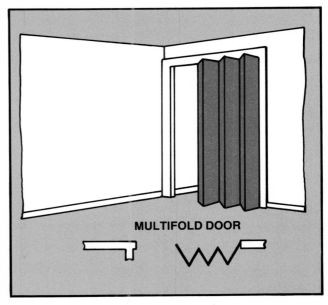

Figure 5.5 Multifolding doors may traverse large areas as a movable partition.

Revolving Doors

Revolving doors present an obstruction to firefighters. Unless they are first collapsed or forced in, they do not permit passage of firefighters in personal protective equipment nor will they allow the passage of hoselines (Figure 5.6a). These doors revolve on central pivots at the top and bottom and usually have four wings at right angles to one another. In some door styles only two wings may be found, each of which has a curved extension piece. The wings on the doors are generally constructed to collapse and move to one side so as to provide a relatively unobstructed opening. There are two methods of securing the wings, as shown in Figure 5.6b.

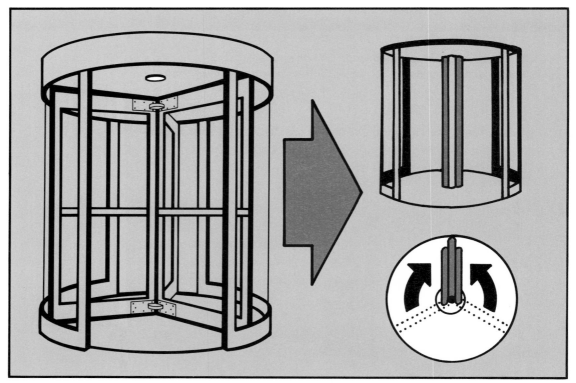

Figure 5.6a Revolving doors will need to be collapsed during emergency operations.

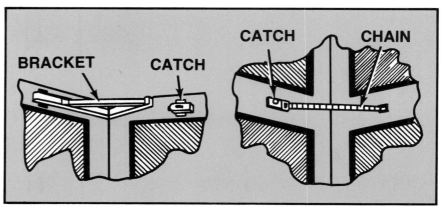

Figure 5.6b Two common ways of securing the wings of a revolving door are a stretcher bar or a chain.

Rolling Doors

Rolling steel doors are designed to be used as both service and fire doors for closing a wall opening at any time. This style of door has its housing and mechanism located at the head of the opening and are composed of a curtain of interlocking metal slats that coil up on a barrel. The barrel is provided with a torsion-spring mechanism to counterbalance the weight of the curtain. These doors may be operated by hand, chain, crank, or electrical power (Figure 5.7).

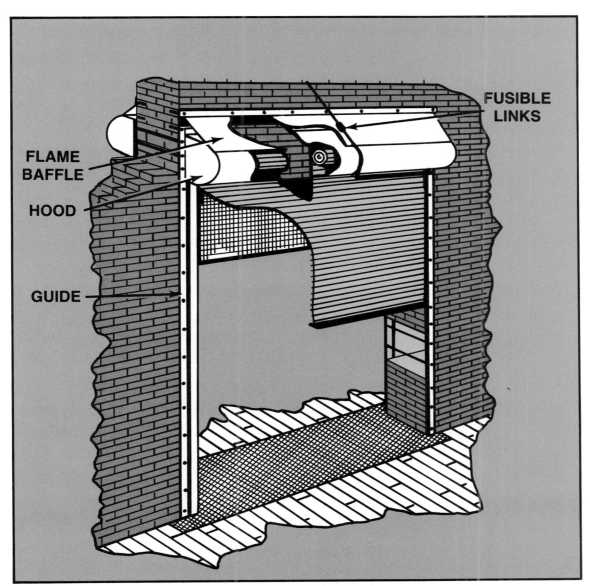

Figure 5.7 Rolling steel doors are often used in fire walls to provide automatic protection.

Rolling doors are equipped with a mechanism that closes the door automatically from any position upon release of the fusible link. One or more torsion springs provide counterbalancing for normal use, and under fire conditions, drive the door to the closed position.

FIRE DOORS

Doorways and other openings in fire walls that separate areas or enclose hazardous materials or processes weaken the wall as a fire barrier and require the best protection available. Therefore, rated fire doors are installed in these openings. A rated fire door is a door that has been awarded an hourly fire rating by passing the American Society for Testing and Materials (ASTM) test E152, *Fire Test of Door Assemblies*. These tests are usually conducted by nationally recognized laboratories; most notable, the Underwriters Laboratories. Underwriters Laboratories (UL) administer the test and if the door meets the criteria set forth in ASTM test 152, the door is awarded a UL label and listing in the appropriate UL listing manuals.

Methods of Operation

SLIDING FIRE DOORS ON TRACKS

Sliding fire doors on tracks are normally balanced by a counterweight for ease of operation. Either a single door to one side of the opening or two doors (one to each side of the opening) may be used. The closing of the door is accomplished by gravity when the fusible link melts and releases the counterweight. The single-slide installation is most commonly used because it is simple in construction and operation, easily maintained, and requires a minimum of floor space (Figure 5.8).

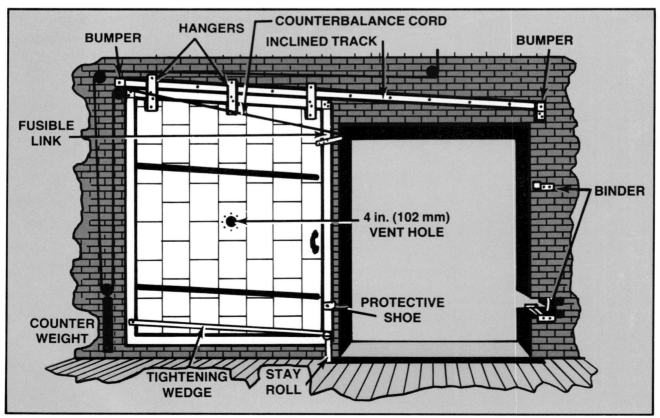

Figure 5.8 Sliding fire doors are a single metal clad door on an inclined track.

SWINGING FIRE DOORS

Swinging fire doors are designed to operate in conjunction with fusible-link-operated door closers or magnetic door holders (Figure 5.9). The latter device is designed to release the door upon

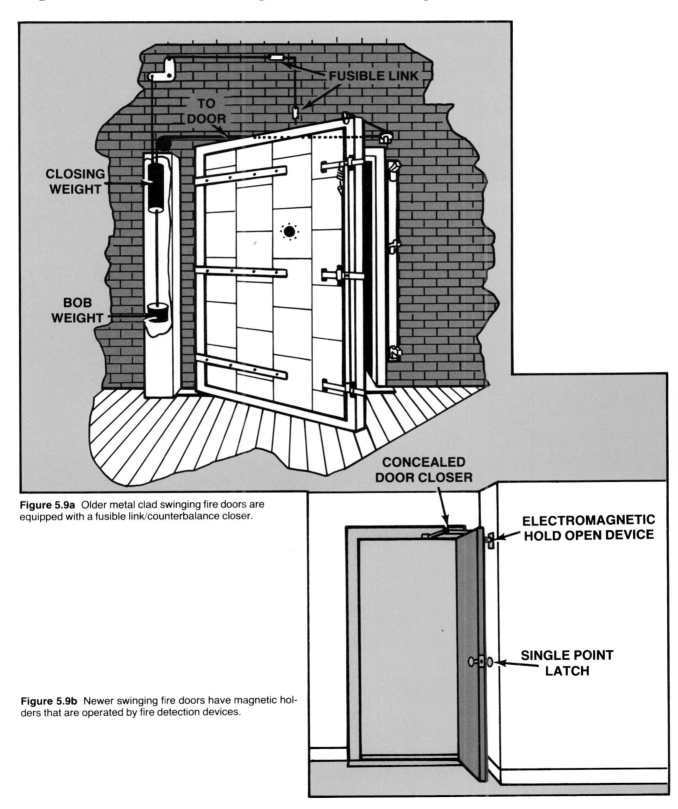

Figure 5.9a Older metal clad swinging fire doors are equipped with a fusible link/counterbalance closer.

Figure 5.9b Newer swinging fire doors have magnetic holders that are operated by fire detection devices.

interruption of an electrical circuit. The interruption may be by approved smoke detectors, devices responsive to heat or smoke, or the loss of electrical power.

Doors designed to swing in pairs are equipped with a mullion that will permit the active leaf to close last. Hardware provisions for these double openings must include appropriate latching devices to maintain the doors in a closed position. Various types of approved panic hardware, either flush or surface-mounted, are available for the application.

TELESCOPING VERTICALLY SLIDING DOORS

The telescoping vertically sliding door is divided horizontally into two leaves, and when opening, the lower leaf overlaps the upper leaf. A large and a small counterweight are attached separately to the lower leaf, which in rising, lifts the upper leaf by means of an additional set of cables (Figure 5.10). The counterweights together keep the door in balance in all positions. Fusing of the link disconnects the small counterweight permitting both leaves to close. The size of the large counterweight controls the speed of closing.

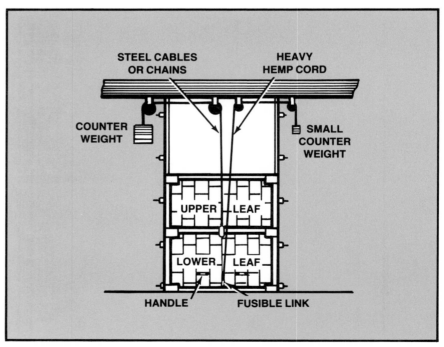

Figure 5.10 Telescoping vertically sliding fire doors are often used for freight elevators.

TYPICAL DOOR CONSTRUCTION

Both listed and nonlisted doors are constructed from a variety of materials. The type of construction and its materials will effectively inform the experienced forcible entry team of the methods necessary to gain entry. This section will address the most common types of door construction.

Solid Core Construction

Solid core construction increases the dimensional stability of the door, provides thermal insulation, and may add fire resistance. Solid core doors are commonly used as exterior and entrance doors. In some applications, doors with a specific fire resistive rating may be required, such as the doors for apartment corridors. These doors must be listed from an approved testing laboratory, such as Underwriters Laboratories or Factory Mutual (Figure 5.11).

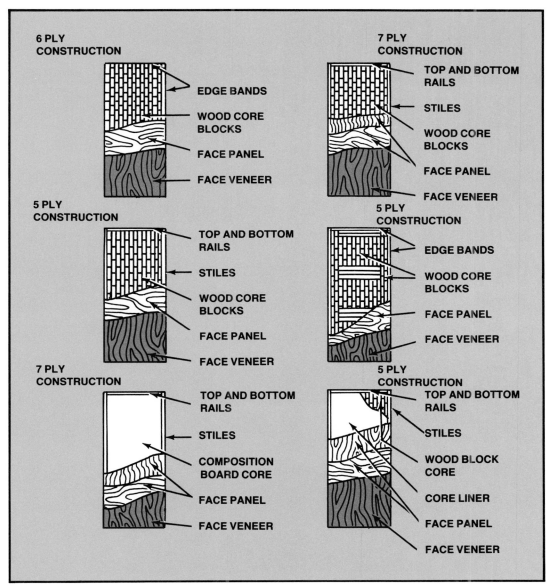

Figure 5.11 Solid core construction for flush doors may be glued block, framed glued block, framed nonglued block, stile and rail, mat-formed composition board, or wood block with a lined core.

Hollow Core Construction

Hollow core door construction has its greatest use in the single-family dwelling as an entrance door into the sleeping areas. Its design provides a structurally sound, yet lightweight door which suits most residential applications. See Figure 5.12.

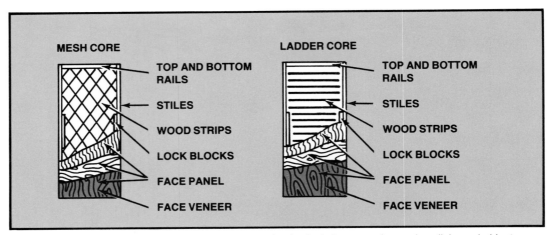

Figure 5.12 Hollow core construction for flush doors may be mesh, cellular, or ladder type.

Hollow Metal Construction

This style of construction utilizes either seamless, flush, panel, rail-and-panel, or style-and-panel design. They are manufactured from a reinforced 20-gage minimum or outer skin supplemented with insulation.

Metal Clad (Kalamein) Construction

Doors which are of this construction style have a metal-covered wood frame provided with insulation or wired-glass panels. The insulated panels consist of composition asbestos beamed ¼-inch (6 mm) or thicker, covered with 24-gage or lighter steel.

Tin Clad Construction

Tin clad doors are made with either a two- or three-ply core of well-seasoned wood covered with lock jointed terneplate or 30-gage sheet metal. The metal covering is nailed to the core.

Sheet Metal Construction

Doors which utilize this style of construction are of the corrugated or flush design, consisting of steel sheets riveted to a steel frame. The corrugated doors have two face sheets of galvanized corrugated steel, with the corrugations at right angles to each other; the face sheets are separated by a layer of asbestos paper, or the hollow spaces between the vertical stiffeners may be filled with loose asbestos. The flush type is covered on both faces with sheets of flat steel.

WINDOW ASSEMBLIES

All windows are classified into the general categories of being fixed (nonoperable), movable (operable) or a combination of fixed and movable. The design and function of each window usually determines the type of installation in any given structure. For example, the fixed will admit light, and depending on the window glazing material, permit or prevent observation through the glass. Economically, the fixed window is the least expensive to install and maintain because of the absence of moving parts. Though not allowed in the sleeping areas of residential dwellings by most model building codes, fixed windows have been found in these types of settings, thus reducing the occupant's chance of survival. Access through a fixed window during a rescue will result in damage to the unit.

Movable units that are properly installed and maintained provide a ventilation channel, and if of sufficient size, afford access to the firefighter. Therefore, it is imperative that firefighters have a sound understanding of window components and their individual functions to minimize window damage when operating on the fireground.

Window Components

A complete window unit consists of a frame, one or more sashes, and all necessary hardware to make a complete operating unit. Window construction will be affected by the way in which a part or the whole of a window is arranged to open.

The term "sash" refers to the movable assembly that holds the glass elements inside the window frame. The sash is composed of horizontal *rails*, vertical *stiles*, and in some instances, *muntins* and *bars* (Figure 5.13). Each of these pieces form the total assem-

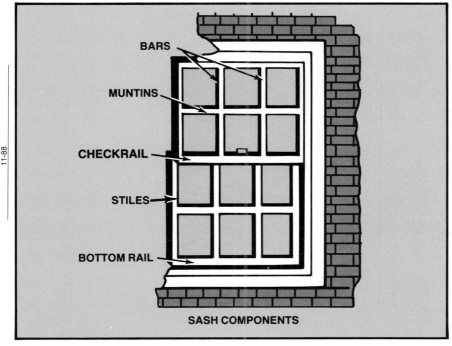

11-88

Figure 5.13 The components of a sash or movable assembly that holds the panes of glass.

bly that allows for the installation of either single or multiple panes of glass, depending upon the owner's and architect's design.

The enclosure for the sash is defined as the *frame* (Figure 5.14). This "housing" fits in the wall opening and is the guide mechanism of the sashes. The frame is composed of the *sill, side jamb,* and *head jamb* (Figure 5.15). The sill is the lowest horizontal member of the window frame and supports the weight of the hardware and sash.

Figure 5.14 The window frame is the enclosure for the window sash or sashes.

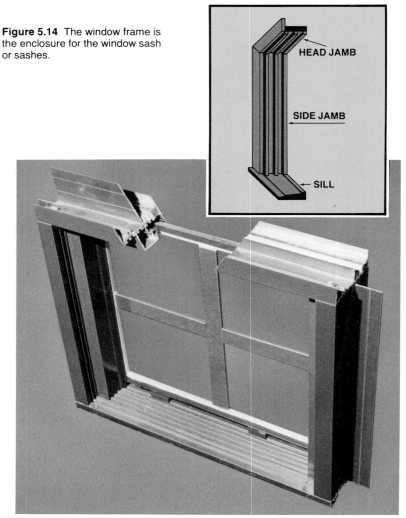

HEAD JAMB

SIDE JAMB

SILL

Figure 5.15 The frame is composed of a sill, side jambs, and head jamb.

Frames are manufactured from treated wood, steel, or aluminum extrusions. Metal, which has greater impact resistance, is usually difficult to form. Wood, on the other hand, has some flexibility, allowing greater ease when forcing entry. The beam surfaces of the frame and sash are susceptible to binding due to moisture, which causes wood to swell or paint overspray. Also, some manufacturers are incorporating plastic or metal guides at the point where sashes and frames meet. Permanent, nonpainted, lubricated slides are excellent for the owner, but can be a burden for the firefighter. Under intense heat the plastic melts, thus rendering the window useless.

Glazing refers to the glass or plastic pane held in the sash with glass points, clips, or wood filler strips. Each sash may have one single pane or a number of panes subdivided by muntins and bars. Sashes with multiple panes are defined as colonial sashes.

Glazing material is manufactured from either silicon glass or plastic. The size and thickness will depend on the size of the window's sash and the window's application. For residential occupancies, windows are commonly single or double glaze (Figure 5.16). Commercial occupancies usually install tempered or plate glass that provides greater strength. Tempered glass is required in new installations near entrances.

Plastics, most commonly Lexan®, are used in areas where security is a concern. The material is resistant to vandalism and is extremely strong. It is also very difficult to force. Usually, power saws are required to cut through the material. For more information see the IFSTA **Forcible Entry** manual.

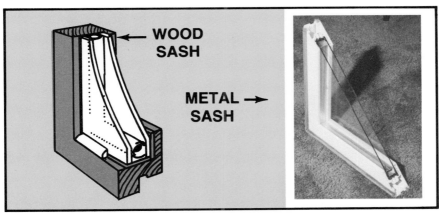

WOOD SASH

METAL → SASH

Figure 5.16 Double glaze windows provide an insulating air space to reduce internal temperature changes.

Sash Balances

There are a large number of windows that utilize balances or tension mechanisms.

Balancing devices for double hung windows are of the following types:

- Spiral spring
- Spring
- Coiled tape
- Counterweight

Balancing hardware is designed to assist in raising the sash and holding it stationary in the desired position.

As shown in Figure 5.17, each balance has a simple but unique operating characteristic. The spiral spring balance operates on a helical principle. The spring guide supports the tension of the helix, which is installed against the sash. Whereas, the spring balance utilizes tension as its fulcrum mechanism. The coiled tape utilizes a compression spring to support the sash, while the counterweight utilizes a balanced weight to support the weight of the sash. Most vertically sliding windows in older construction use a counterweight. Since the counterweight moves in an open vertical channel beside or outside the frame, fire can spread rapidly through it. As a result, the window casing must be removed during overhaul.

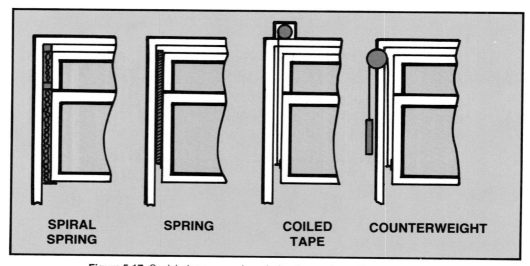

SPIRAL SPRING   **SPRING**   **COILED TAPE**   **COUNTERWEIGHT**

Figure 5.17 Sash balances may be spiral spring, spring, coiled tape, or counterweight in operation.

TYPES OF WINDOWS

Fixed Windows

Fixed windows usually consist of a frame and glazed stationary sash (Figure 5.18). They are often flanked with double hung and casement windows, or stacked with awning and hopper units to make up windows of custom designs. Unless the glass pane is destroyed, none of the window area is available for ventilation.

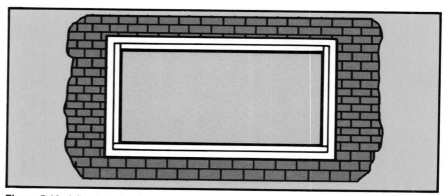

Figure 5.18 A fixed window has a sash that does not open.

Double and Single Hung Windows

Double hung windows have two operating sashes: single hung windows have only the lower sash operative (Figure 5.19). The sashes move vertically within the window frame and are maintained in the desired position by friction fit against the frame channels or with balancing devices. Balancing devices also assist in raising the sash. Fifty percent of the window's area is available for ventilation.

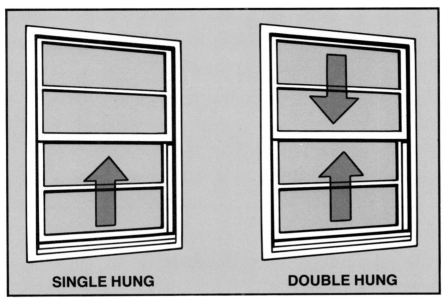

SINGLE HUNG   DOUBLE HUNG

Figure 5.19 Single and double hung windows have an operating sash for ventilation.

Casement Windows

Casement windows have side-hinged sash generally installed to swing outward (Figure 5.20). They may contain one or two operating sashes and sometimes a fixed light between the pair of

Figure 5.20 Casement windows open on one side, usually outward, with the screen on the inside.

sash. A fixed light is a nonopening window. When fixed lights are not used, a pair of casements may close on a mullion or against themselves, providing an unobstructed view when open. One hundred percent of the window is available for ventilation.

Horizontal Sliding Windows

Horizontal sliding windows have two or more sashes of which at least one moves horizontally within the window frame (Figure 5.21). In a three-sash design, the middle sash is usually fixed; in two-sash units, one or both sashes may be ventilating. Ventilating area is 50 percent of the window area in most designs.

Figure 5.21 Horizontal sliding windows have sashes that move to the side.

Awning Windows

Awning windows have one or more top-hinged, outswinging sashes (Figure 5.22). Single awning sash is often combined with fixed and other types of sashes into larger window units. Ventilation area is considered to be 100 percent of the operating sash areas.

Figure 5.22 Awning windows have a sash that is hinged at the top.

Jalousie Windows

Jalousie windows consist of small glass sections about four-inches (102 mm) wide and as long as the window width (Figure 5.23). The window sections are usually without sashes and the glass is heavy plate that has been ground to overlap when closed. These glass sections are supported on each end by a metal operating mechanism. This mechanism may be exposed or concealed along the sides of the window, and each glass panel opens the same distance outward when the crank is turned. The operating crank and gear housing are located at the bottom of the window. Ventilation area is considered to be 100 percent of the operating sash areas.

Figure 5.23 Jalousie windows are of a heavy ground glass without a sash.

Projected Windows

Projected or factory windows are ordinarily made of metal and they may project in or out from an opening (Figure 5.24). They may pivot at the top or bottom. "Projected-Out" factory windows swing outward at the bottom and slide down from the top in

Figure 5.24 Projected or factory windows may project in or out at the bottom.

a groove which is provided for that purpose. "Projected-In" factory windows swing inward at the top and they are usually hinged at the bottom. Pivoted projected windows are usually operated by a push bar that is notched to hold the window in place. In some cases, banks of projected windows are opened mechanically. Screens are seldom used with this type of window, but when they are, they are on the side opposite the direction of projection.

Hopper Windows

Hopper windows have one or more bottom-hinged, in-swinging sashes (Figure 5.25). Hopper sash is similar in design and operation to the awning type and may actually be an inverted awning sash with minor hardware and weather stripping modifications. For this reason, windows with hopper sash sometimes are referred to as awning windows. Operating sash provides 100 percent ventilation area.

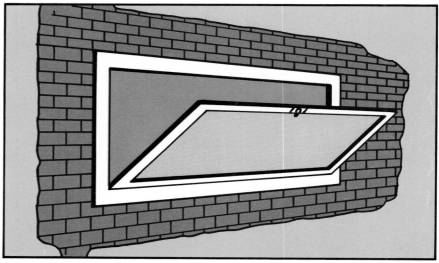

Figure 5.25 Hopper windows are similar to awning windows only hinged at the bottom and swing inward.

Pivoted Windows

Pivot windows are windows that utilize one or more sashes that are hinged or pivoted in either a horizontal or vertical plane. The pivot or hinge points are located in the center of the window frame and are designed so that part of the sash opens inward and part outward. Figure 5.26 shows both a horizontal and vertical pivot window.

Pivot windows provide poor escape access for building occupants and are not allowed by most model building codes in bedrooms or residential occupancies. However, because of their design, this window style provides a 100 percent ventilation opening.

Locking devices for this style of window are usually located at either the top or bottom of the window frame or the pivot points.

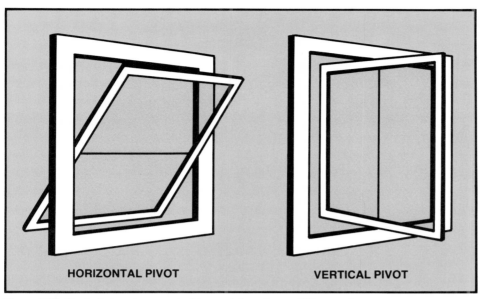

HORIZONTAL PIVOT   **VERTICAL PIVOT**

Figure 5.26 Pivot windows have center hinge points and open 100 percent.

Security Windows

As the name implies, the security window is designed to prevent unauthorized entry and egress of an occupancy, either for security or safety reasons. Metal bars are fastened to the exterior of the window frame for this purpose. These bars are usually constructed of minimum ¼-inch (6.2 mm) steel that is welded into an assembly for ease of installation (Figure 5.27). This style of window may incorporate either fixed or movable sashes.

While providing access control, this window has a long history of contributing to fire death because it prevents an occupant's ability to escape from a building. For this reason, most municipalities require that new metal bar assemblies be provided with a method for unfastening that is operable from the interior.

Figure 5.27 Security windows are barred and present numerous problems during fire conditions. *Courtesy of Edward Prendergast.*

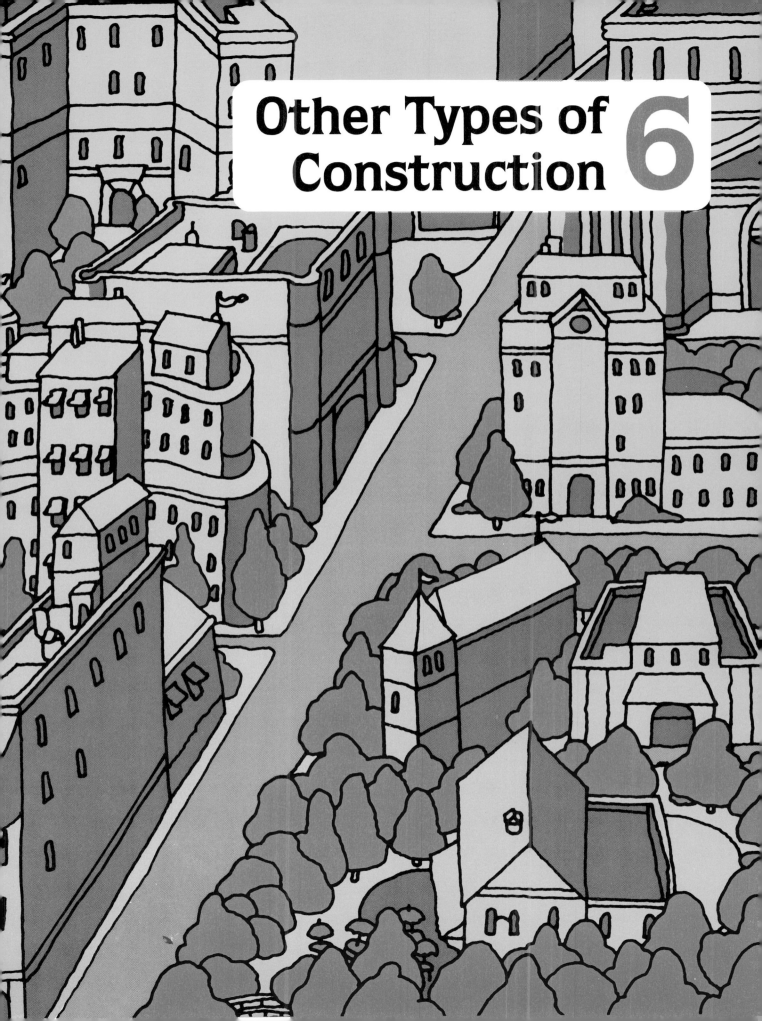

Other Types of Construction 6

Chapter 6
Other Types of Construction

MOBILE HOMES

Mobile homes, prefabricated houses, and modular units are collectively designated, and legally referenced as, "Manufactured Housing." The mobile home market represents the largest segment of manufactured housing and presents the newest, most active developments. Nationwide, mobile homes account for 37 percent (approximately 240,000 units) of new single-family housing in 1982. Manufacturers shipped approximately 65,000 new manufactured houses in the first quarter of 1983.

The image of the mobile home has long been that of transients, junked automobiles, and cramped trailers listing on cinder block stilts. Today's reality does not support the historical perception of eccentricity. Current editions, though movable, are most often permanently anchored to concrete slabs. State-of-the-art production provides living space equivalent to many new homes and offers an array of built-in luxuries. An available model might be paneled, furnished, and have bay windows, redwood decks, sunken bath tubs, and wet bars.

Unlike traditional mobile homes, homes built after June 15, 1976 are constructed according to guidelines of the United States Department of Housing and Urban Development (HUD) Standard 24 CFR, Chapter 14, the *Federal Mobile Homes Construction and Safety Standard* (FMHCSS). The FMHCSS is a performance building code, rather than the prescriptive building code used by most municipal entities. The efficiency realized from this type of construction explains the lower price of the homes. Mobile homes built to this standard must display a seal that indicates conformity to HUD criteria. Mobile homes built before the aforementioned date must be built to NFPA Standard No. 501 A, *Mobile Home Installations, Sites, and Communities* in addition to the standards of the American National Standard Institute (ANSI).

Mobile Home Construction

The typical mobile home is an assembly of four major components: the chassis, floor, wall, and roof systems. Although they are constructed of steel, wood, plywood, aluminum, and gypsum wallboard, among other materials, they are basically frame construction. About 70 percent of all mobile homes conform to general design: single section units (single-wide) are typically 14 feet (4.3 m) wide (the maximum allowable width on interstate highways), and between 55 feet and 70 feet (17 m and 21 m) in length (Figure 6.1). The remaining 30 percent are comprised of multisections (multiwide): two or more sections that when set on a permanent foundation can appear indistinguishable from houses of other construction (Figure 6.2). Sizes range from 24 feet to 28 feet (7.4 m to 8.5 m) wide, and from 60 feet to 70 feet (18 m to 21 m)

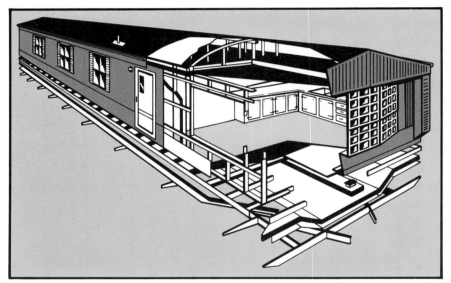

Figure 6.1 An exploded view of a single-section mobile home.

Figure 6.2 Double-wide mobile homes are shipped in two sections and secured together on location.

long. A crawl space is an option, as is a basement. Useable living space ranges from 966 square feet (89.7 m^2) for a typical single-wide, to 1,440 square feet (133.7 m^2) for a typical double-wide mobile home.

ROOF CONSTRUCTION

The roof construction of most newer mobile homes uses the shallow bowstring truss for structural support. Even though these trusses are designed to support a minimum load of 1,000 pounds (453.6 kg) per pair, when placed in a conventional supporting method, their integrity should always be monitored. This type of construction forms a void between the roof material and the interior ceiling allowing an avenue for fire extension. Thirty gage sheet metal is the most common material used for roofs, while low density, Class C fire-resistive fiberboard is used as the interior ceiling material (Figure 6.3). Therefore, it is imperative that the interior ceiling be opened during fire fighting operations to allow for an investigation of possible fire communication.

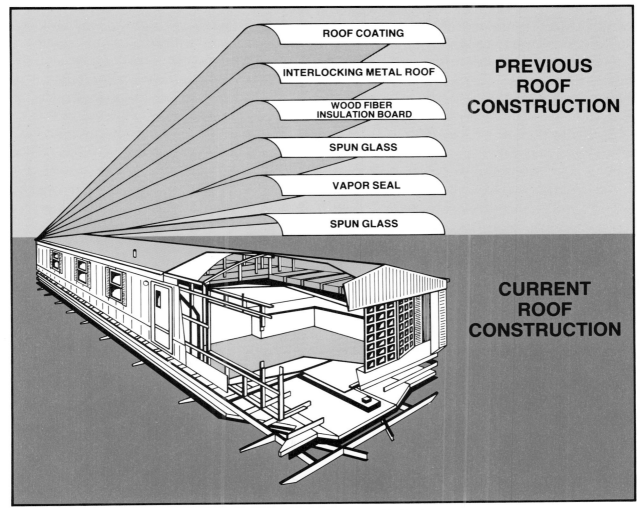

Figure 6.3 Details of roof support, roof covering, and floor construction are provided to show possible avenues of fire spread.

Though roof ventilation would be beneficial, the integrity of mobile home roofs during a fire is questionable. Therefore, vertical ventilation cannot always be accomplished.

WALL CONSTRUCTION

Wall construction usually consists of nonfire-stopped 2x4 inch (51 mm by 102 mm) wood construction with some diagonal bracing for increased wind resistance. Most mobile homes have wall studs located on 36-inch (914 mm) centers. The walls are designed to support a horizontal load of not less than 5 pounds per square foot (2.4 kg/m²). From the floor, the usual wall heights range from 6½ feet to 8 feet (1.9 m to 2.4 m). Interior walls are usually prefinished plywood paneling with a 5/16-inch (8 mm) gypsum board backing. FMHCSS regulations require that this interior finish must have a minimum Class C rating (Figure 6.4).

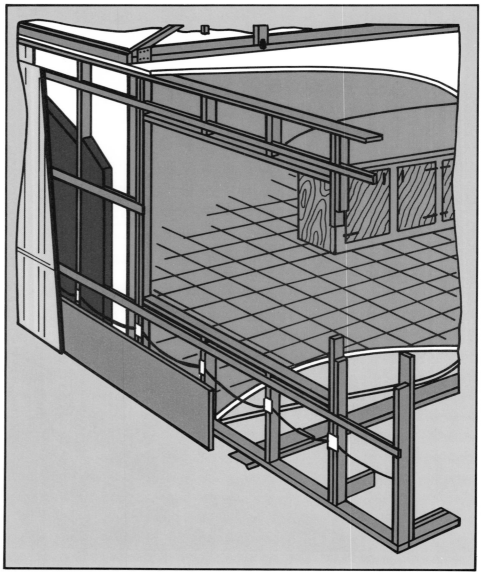

Figure 6.4 Mobile home walls are of lightweight construction and nonfire-stopped.

FIRE FIGHTING CONSIDERATIONS

The code produced manufactured housing unit is a structure clad with metal on all four sides, and in most cases, covered with a metal roof. Such a structure is capable of containing smoke and heat efficiently, allowing a rapid build-up of superheated fire gases. This can result in flashover. To prevent this, either vertical or horizontal ventilation is necessary. The majority of mobile homes have long, narrow hallways. This, in conjunction with the low ceiling height, promotes a rapid fire extension. Plastics are also being used due to the savings in both cost and weight. The most prominently used plastics are polyvinyl chloride (PVC) and polyvinyl dichloride (CPVC). Both are respiratory irritants and are possible carcinogens. Therefore, it is imperative that positive-pressure self-contained breathing apparatus is used throughout fire fighting operations.

Bedroom windows afford a quick efficient means for the removal of trapped occupants. The FMHCSS requires that any sleeping area that does not have a door leading to the exterior be equipped with a window with an area equal to or greater than five square feet ($.5 \text{ m}^2$).

Problems can also arise from the exterior finish and aesthetics. Many of the newer homes have a plywood exterior finish which can cause exposure problems. Also, skirting is often required by mobile home court operators to hide the wheels. If these areas are not policed, many home owners will use this area for the storage of combustible and flammable materials.

Mobile home courts or parks can range from the poorly managed and maintained to the elite and professionally maintained. Older parks generally were designed for smaller trailers. Consequently, they become quite crowded when newer, larger mobile homes are moved in. This results in reduced space between homes and increased difficulties when controlling exposures. In addition, the population density is usually high, creating both rescue and crowd management problems. Most mobile home communities are zoned on land that is not in demand for more lucrative purposes. Generally, they are located on the perimeter to the community, resulting in increased response times. In some cases, they do not have municipal water available.

PREFABRICATED CONSTRUCTION

Prefabricated or panelized construction defines a method of building construction where the walls, floors, and ceilings are manufactured complete with plumbing, electrical wiring, and all mill work. Once delivered to the site, the entire assembly is erected. Being lightweight in construction promotes a quick erection time. Though structurally sound, this form of manufactured

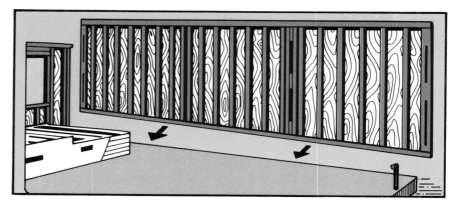

Figure 6.5 Prefabricated housing uses prebuilt panels for rapid construction.

construction is generally associated with rapid fire spread due to the number of common voids (Figure 6.5).

MODULAR CONSTRUCTION

A modular building is built in two or more sections at the factory (Figure 6.6). All utilities such as plumbing, heating, and electrical systems are installed as an integral part of the con-

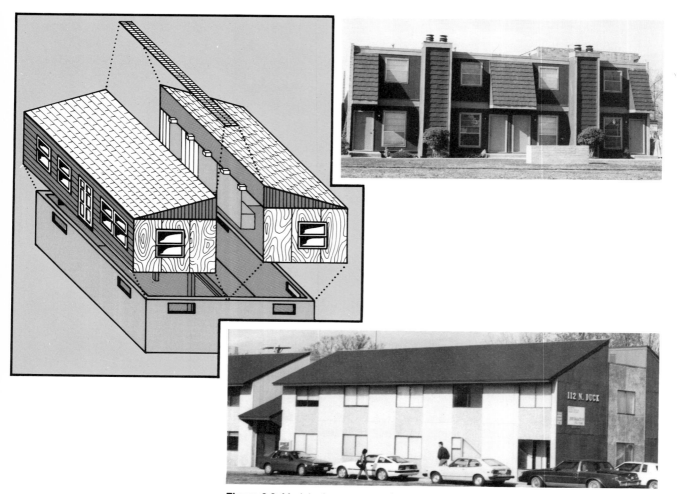

Figure 6.6 Modular homes are entire sections completed at the factory and transported to a permanent location for assembly.

struction phase. In addition, all mill work such as doors and windows are installed by the manufacturer. The modular unit is designed to conform to the building codes of the municipality where the building is to be located. Some modulars are designed to be stacked together to form multiple-story buildings. As in most manufactured buildings, lightweight components are used. The components are structurally sound as long as they are not attacked by fire. Breaching of any structural member in this type of construction can promote structural collapse.

GEODESIC DOME CONSTRUCTION

A geodesic structure is defined as a dome or vault made of lightweight straight structural elements installed so as to form a tension load (Figure 6.7). The principle of geodesic construction is to reduce the weight at the tension points so as to make the structure economical to build. But being domed or hemispherical in shape promotes a greater rate of fire spread, and difficulty in laddering the structure. These factors can lead to a rapid degradation of structural stability resulting in structural collapse.

Figure 6.7 Geodesic domes are of lightweight construction, but can fail quickly under fire conditions.

LOG HOMES

Log homes or "cabins" are another style of premanufactured housing. These homes are constructed from solid logs ranging from four inches to nine inches (101 mm to 228 mm) in diameter. The ends of the logs are machined to form a tight, structurally sound fit (Figure 6.8). Because of their large surface area, log homes will contain a great amount of heat that can lead to a back-draft if ventilation is not completed promptly. In addition, some homeowners apply a coating of polyurethane varnish to maintain the rustic appearance. This material, when applied on vertical surfaces, will increase the rapidity of the flame spread.

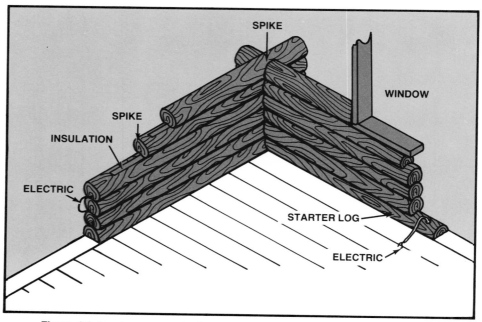

Figure 6.8 Log homes are simple to construct and do not have hidden channels for fire spread.

CONSTRUCTION MATERIALS

Most of the materials utilized in the construction of manufactured housing have been addressed in other sections of this manual. However, the following list addresses some of the more common materials found within manufactured housing.

- Cement or Cinder Blocks: Constructed in a square shape with rectangular openings, these blocks are used as foundations and walls. They can be used as a structural component or as a decorative feature.

- Trusses: Trusses are the supporting mechanism for the roof assembly of most styles of manufactured housing. Manufactured from wood, steel, or a combination thereof, they are the most prominent design used. Though structurally sound, they are known to perform poorly in a fire situation.

- Prestressed concrete: Utilizing reinforcement bars or "rebar," this type of concrete is extremely strong. The material has excellent fire-resistive features and will remain structurally sound even after a severe fire where spalling was present.

AGRICULTURAL-TYPE BUILDINGS

Buildings used in an agricultural setting are usually large buildings with a greater than average square footage (m²). These large interior areas are used for the storage of farm machinery, silage, or foliage. Being located in a rural setting, these buildings will often contain the primary assets of the farm; consequently, fire protection should be a concern. Statistics have shown, however, that this is not always the case. If a fire is reported in a building located ten minutes away from the first due companies (which is often the case), firefighters should expect to find a steadily increasing working structure fire.

Agricultural-type buildings are not limited to agricultural or farm environments. The type of construction includes, in general, a more skeletal framework of perimeter walls and roofs with few internal partitions. Ag-type buildings may be totally enclosed, insulated, and otherwise finished depending on intended use.

To assist the firefighter in combating a fire in this setting, this section will address the two most common types of building construction in the agricultural setting: pole-barn- type construction and quonset-type construction.

Pole-Type Construction

Pole-type construction utilizes round or sawed posts set directly in the ground. Installed with a concrete footing, the poles serve as bearing wall supports and with the addition of girts, purlins, rafters, or trusses, the assembly forms a complete frame (Figure 6.9). With the attachment of girders to the floor pole

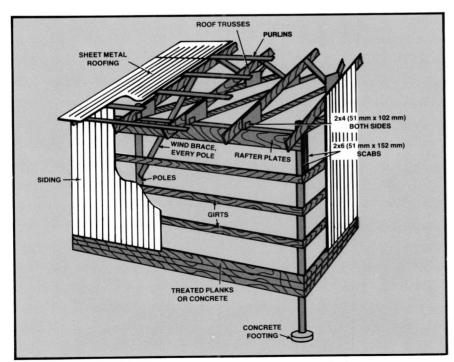

Figure 6.9 Pole-wall-type buildings are common in agricultural applications.

joists, finished floors can be constructed. Thus, homes could be built utilizing pole construction if local codes permit.

Attached to poles are structure members called girts. Girts are used to reinforce the rigidity of the poles and to allow for the attachment of siding (Figure 6.10).

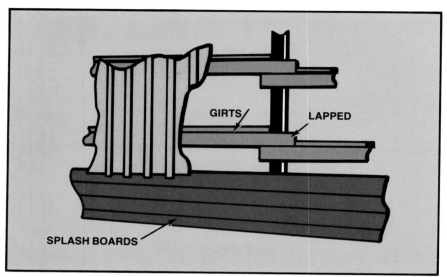

Figure 6.10 Girts run horizontally in the walls to provide added support.

Floor construction is usually concrete, though it is not uncommon to find dirt or gravel as the flooring material.

Roof construction is usually truss or rafter type and is installed by resting the assembly on top of the poles. The most common type of roof material is metal sheets, either stapled or nailed in place (Figure 6.11).

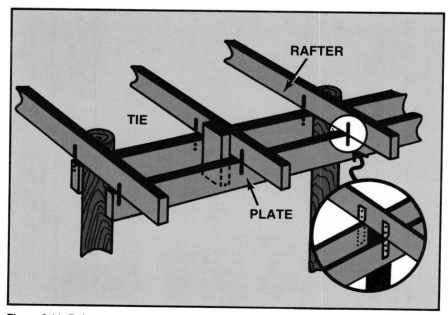

Figure 6.11 Rafters or trusses secured to the poles are used to support the roof coverings.

Stud Wall Construction

Stud wall construction for agricultural buildings is similar to ordinary frame construction. Wall plates and studs are usually on 24-inch (610 mm) center versus 16-inch (406 mm) for residential construction. The exterior siding is usually metal, plywood, or tongue and groove planks. Foundation construction is commonly masonry but can be heavy timber.

Rigid Frame Construction

Rigid frame construction utilizes straight wall stud and roof rafters to form a unitized wall component. Once erected, this type of construction allows for a completely open interior space. Frame members are anchored to either a wood sill or to a concrete foundation. Either metal sachets bolted to plates or anchor belts are used.

Frame members consist of solid wood studs with a truss rafter, or a steel frame assembly (Figure 6.12). A variation of wood construction is the use of a truss frame covered by plywood. Wall coverings are usually plywood, formed metal, or corrugated metal installed directly to the frame members.

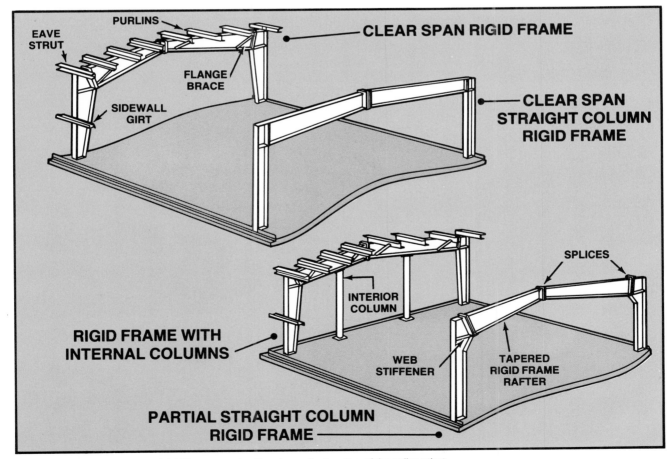

Figure 6.12 This metal bent of rigid frame construction shows the location of the splices that join the various components together and where the webs have been stiffened by the addition of welded plates to strengthen the frame and prevent web buckling.

Quonset Buildings

Quonset huts were developed during World War II to provide a quick, easily erected temporary structure. The prefabricated structure consists of curved corrugated metal panels formed to construct a continuous curve (Figure 6.13). This forms the wall and roof into one complete assembly. Being panelized in design, the building may be of great length. Ends are enclosed with vertical panels with openings for doors and windows.

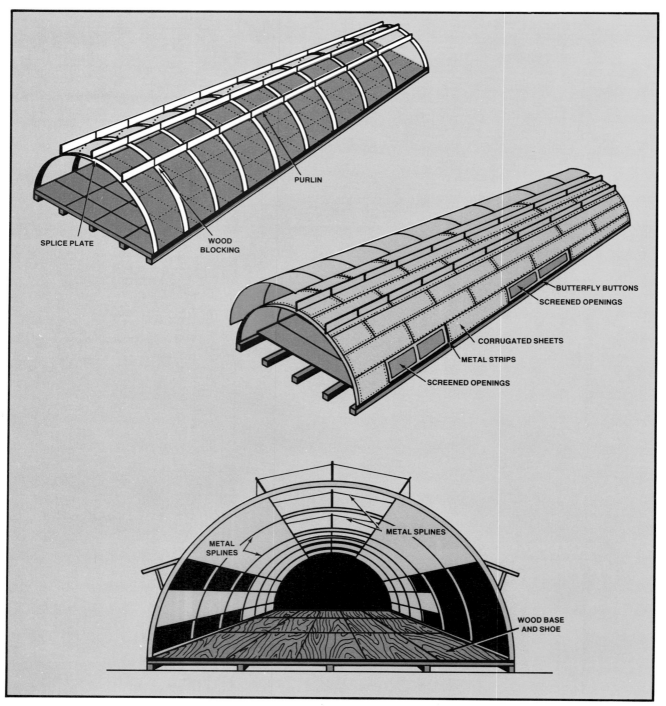

Figure 6.13 Quonset buildings have the wall and roof as one integral assembly.

EARTH SHELTERED HOUSING

In the mid 1960s an idea was conceived of integrating earthen enclosed structures with solar power. This development has evolved into a practical, efficient, and economical form of housing called earth sheltered housing. The two basic types of earth sheltered housing are the berm and the excavation (Figure 6.14). The berm style of construction is utilized when the structure is built above ground and covered with earth; the excavation style is built below grade and then recovered with earth.

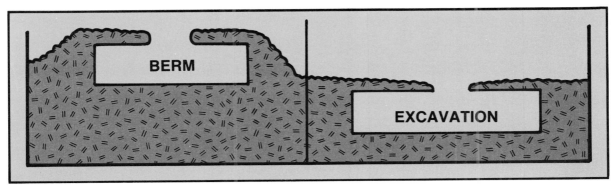

Figure 6.14a The various styles of earth sheltered homes are built around the berm and excavation principles.

Figure 6.14b As in conventional residential construction, there are a variety of differing styles of earth sheltered housing which the firefighter may encounter.

Several considerations and problems are associated with earth sheltered housing. These include:

- Geographic location and orientation
  — Most earth sheltered homes are located in a rural setting; consequently, fire department response time is usually increased.

- Structure design
  — Since dirt is constantly creating force on the structural members, the chances of collapse can be greater if structure members degrade due to fire attack.

Other areas of concern include drainage and ventilation, water resistance of walls, and provisions for solar heating.

Firefighters should not assume that a fire in an earth sheltered home will be the same as a basement fire, though similarities do exist. The primary concern is that of limited entry; access is usually limited to the front door and openings for solar heating. Usually these entrances are sloped, thereby making approach difficult.

Options for ventilation are limited. Because the number of openings is limited, difficulty will be encountered when attempting to incorporate natural or forced ventilation.

There are numerous designs and variations of earth sheltered structures that preclude any generalized approach to tactics. Each earth sheltered structure must be surveyed and appropriate tactics and options should be developed.

Index

IFSTA MANUALS AND FPP PRODUCTS

For a current catalog describing these and other products, call or write your local IFSTA distributor or Fire Protection Publications, IFSTA Headquarters, Oklahoma State University, Stillwater, OK 74078-0118.
Phone: 1-800-654-4055

Awareness Level Training for Hazardous Materials
prepares fire, police, EMS, and public utilities to recognize and identify the presence of hazardous materials at an emergency scene. Addresses the requirements in NFPA 472, Chapter 2: Competencies for First Responders at the Awareness Level. Information found in the manual includes responsibilities of the first responder, identification systems, types of containers, and personal protective equipment. 1st Edition (1995), 152 pages.

Study Guide Awareness Level Training For Hazardous Materials
The companion study guide in question and answer format. (1995), 184 pages.

Fire Department AERIAL APPARATUS
includes information on the driver/operator's qualifications; vehicle operation; types of aerial apparatus; positioning, stabilizing, and operating aerial devices; tactics for aerial devices; and maintaining, testing, and purchasing aerial apparatus. Detailed appendices describe specific manufacturers' aerial devices. 1st Edition (1991), 386 pages, addresses NFPA 1002.

STUDY GUIDE FOR AERIAL APPARATUS
The companion study guide in question and answer format. 1991, 140 pages.

AIRCRAFT RESCUE AND FIRE FIGHTING
comprehensively covers commercial, military, and general aviation. All the information you need is in one place. Subjects covered include: personal protective equipment, apparatus and equipment, extinguishing agents, engines and systems, fire fighting procedures, hazardous materials, and fire prevention. Over 240 photographs and two-color illustrations. It also contains a glossary and review questions with answers. 3rd Edition (1992), 247 pages, addresses NFPA 1003.

BUILDING CONSTRUCTION RELATED TO THE FIRE SERVICE
helps firefighters become aware of the many construction designs and features of buildings found in a typical first alarm district and how these designs serve or hinder the suppression effort. Subjects include construction principles, assemblies and their resistance to fire, building services, door and window assemblies, and special types of structures. 1st Edition (1986), 166 pages, addresses NFPA 1001 and NFPA 1031, levels I & II.

CHIEF OFFICER
lists, explains, and illustrates the skills necessary to plan and maintain an efficient and cost-effective fire department. The combination of an ever-increasing fire problem, spiraling personnel and equipment costs, and the development of new technologies and methods for decision making requires far more than expertise in fire suppression. Today's chief officer must possess the ability to plan and administrate as well as have political expertise. 1st Edition (1985), 211 pages, addresses NFPA 1021, level VI.

SELF-INSTRUCTION FOR CHIEF OFFICER
The companion study guide in question and answer format. 1986, 142 pages.

FIRE DEPARTMENT COMPANY OFFICER
focuses on the basic principles of fire department organization, working relationships, and personnel management. For the firefighter aspiring to become a company officer, or a company officer wishing to improve management skills, this manual helps develop and improve the necessary traits to effectively manage the fire company. 2nd Edition (1990), 278 pages, addresses NFPA 1021, levels I, II, & III.

COMPANY OFFICER STUDY GUIDE
The companion study guide in question and answer format. Includes problem applications and case studies. 1991, 243 pages.

ESSENTIALS OF FIRE FIGHTING
is the "bible" on basic firefighter skills and is used throughout the world. The easy-to-read format is enhanced by 1,600 photographs and illustrations. Step-by-step instructions are provided for many fire fighting tasks. Topics covered include: personal protective equipment, building construction, firefighter safety, fire behavior, portable extinguishers, SCBA, ropes and knots, rescue, forcible entry, ventilation, communications, water supplies, fire streams, hose, fire cause determination, public fire education and prevention, fire suppression techniques, ladders, salvage and overhaul, and automatic sprinkler systems. 3rd Edition (1992), 590 pages, addresses NFPA 1001.

STUDY GUIDE FOR 3rd EDITION OF ESSENTIALS OF FIRE FIGHTING
The companion learning tool for the new 3rd edition of the manual. It contains questions and answers to help you learn the important information in the book. 1992, 322 pages.

PRINCIPLES OF EXTRICATION
leads you step-by-step through the procedures for disentangling victims from cars, buses, trains, farm equipment, and industrial situations. Fully illustrated with color diagrams and more than 500 photographs. It includes rescue company organization, protective clothing, and evaluating resources. Review questions with answers at the end of each chapter. 1st Edition (1990), 365 pages.

FIRE CAUSE DETERMINATION
gives you the information necessary to make on-scene fire cause determinations. You will know when to call for a trained investigator, and you will be able to help the investigator. It includes a profile of firesetters, finding origin and cause, documenting evidence, interviewing witnesses, and courtroom demeanor. 1st Edition (1982), 159 pages, addresses NFPA 1021, Fire Officer I, and NFPA 1031, levels I & II.

FIRE SERVICE FIRST RESPONDER
provides the information needed to evaluate and treat patients with serious injuries or illnesses. It familiarizes the reader with a wide variety of medical equipment and supplies. **First Responder**

applies to safety, security, fire brigade, and law enforcement personnel, as well as fire service personnel, who are required to administer emergency medical care. 1st Edition (1987), 340 pages, addresses NFPA 1001, levels I & II, and DOT First Responder.

FORCIBLE ENTRY
reflects the growing concern for the reduction of property damage as well as firefighter safety. This comprehensive manual contains technical information about forcible entry tactics, tools, and methods, as well as door, window, and wall construction. Tactics discuss the degree of danger to the structure and leaving the building secure after entry. Includes a section on locks and through-the-lock entry. Review questions and answers at the end of each chapter. 7th Edition (1987), 270 pages, helpful for NFPA 1001.

GROUND COVER FIRE FIGHTING PRACTICES
explains the dramatic difference between structural fire fighting and wildland fire fighting. Ground cover fires include fires in weeds, grass, field crops, and brush. It discusses the apparatus, equipment, and extinguishing agents used to combat wildland fires. Outdoor fire behavior and how fuels, weather, and topography affect fire spread are explained. The text also covers personnel safety, management, and suppression methods. It contains a glossary, sample fire operation plan, fire control organization system, fire origin and cause determination, and water expansion pump systems. 2nd Edition (1982), 152 pages.

FIRE SERVICE GROUND LADDER PRACTICES
is a "how to" manual for learning how to handle, raise, and climb ground ladders; it also details maintenance and service testing. Basic information is presented with a variety of methods that allow the readers to select the best method for their locale. The chapter on Special Uses includes: ladders as a stretcher, a slide, a float drag, a water chute, and more. The manual contains a glossary, review questions and answers, and a sample testing and repair form. 8th Edition (1984), 388 pages, addresses NFPA 1001.

HAZARDOUS MATERIALS FOR FIRST RESPONDERS
prepares the reader to meet the objectives for First Responder at the Awareness and Operational levels contained in NFPA 472. The manual includes information on properties of hazardous materials, recognizing and identifying hazardous materials, personal protective equipment, emergency scene command and control, incident control tactics and strategies, and decontamination. Over 350 illustrations are used to reinforce the text. 2nd Edition (1994), 241 pages.

STUDY GUIDE FOR IFSTA HAZARDOUS MATERIALS FOR FIRST RESPONDERS
The companion study guide in question and answer format. 2nd Edition (1994), 253 pages.

HAZARDOUS MATERIALS: MANAGING THE INCIDENT
addresses OSHA 1910.120 and NFPA 472, *Standard for Professional Competence of Responders to Hazardous Materials Incidents*. Provides the reader with a logical, systematic process for responding to and managing hazardous materials emergencies. It is directed toward the haz mat technician, incident commander, the off-site specialty employee, and haz mat response team members. Includes numerous charts, diagrams, scan sheets, checklists, and reference information. Topics include haz mat management system, health and safety, ICS, politics of haz mat incident management, hazard and risk evaluation, decontamination, and more! 2nd Edition (1994)

STUDENT WORKBOOK FOR HAZARDOUS MATERIALS: MANAGING THE INCIDENT
The companion study guide in question and answer format. 2nd Edition (1994)

INSTRUCTOR'S GUIDE FOR HAZARDOUS MATERIALS: MANAGING THE INCIDENT
Provides lessons based on each chapter. 2nd Edition (1994)

HAZ MAT RESPONSE TEAM LEAK AND SPILL GUIDE
contains articles by Michael Hildebrand reprinted from *Speaking of Fire's* popular Hazardous Materials Nuts and Bolts series. Two additional articles from *Speaking of Fire* and the hazardous material incident SOP from the Chicago Fire Department are also included. 1st Edition (1984), 57 pages.

EMERGENCY OPERATIONS IN HIGH-RACK STORAGE
is a concise summary of emergency operations in the high-rack storage area of a warehouse. It explains how to develop a pre-emergency plan, what equipment will be necessary to implement the plan, type and amount of training personnel will need to handle an emergency, and interfacing with various agencies. Includes consideration questions, points not to be overlooked, and trial scenarios. 1st Edition (1981), 97 pages.

HOSE PRACTICES
reflects the latest information on modern fire hose and couplings. It is the most comprehensive single source about hose and its use. The manual details basic methods of handling hose, including large diameter hose. It is fully illustrated with photographs showing loads, evolutions, and techniques. This complete and practical book explains the national standards for hose and couplings. 7th Edition (1988), 245 pages, addresses NFPA 1001.

FIRE PROTECTION HYDRAULICS AND WATER SUPPLY ANALYSIS
covers the quantity and pressure of water needed to provide adequate fire protection, the ability of existing water supply systems to provide fire protection, the adequacy of a water supply for a sprinkler system, and alternatives for deficient water supply systems. 1st Edition (1990), 340 pages.

INCIDENT COMMAND SYSTEM (ICS)
was developed by a multiagency task force. Using this system, fire, police, and other government groups can operate together effectively under a single command. The system is modular and can be used to meet the requirements of both day-to-day and large-incident operations. It is the approved basic command system taught at the National Fire Academy. 1st Edition (1983), 220 pages, helpful for NFPA 1021.

INDUSTRIAL FIRE BRIGADE TRAINING: INCIPIENT LEVEL
assists management in complying with applicable laws and regulations, primarily NFPA 600 and 29 CFR 1910, and to assist them in training those who provide incipient level fire protection for industrial occupancies. It is also intended to serve as a reference and training resource for individual emergency responders. 1st Edition (1995), 184 pages.

FIRE INSPECTION AND CODE ENFORCEMENT
provides a comprehensive, state-of-the-art reference and training manual for both uniformed and civilian inspectors. It is a comprehensive guide to the principles and techniques of inspection. Text includes information on how fire travels, electrical hazards, and fire resistance requirements. It covers storage, handling, and use of hazardous materials; fire protection systems; and building construction for fire and life safety. 5th Edition (1987), 316 pages, addresses NFPA 1001 and NFPA 1031, levels I & II.

STUDY GUIDE FOR FIRE INSPECTION AND CODE ENFORCEMENT

The companion study guide in question and answer format with case studies. 1989, 272 pages.

FIRE SERVICE INSTRUCTOR

explains the characteristics of a good instructor, shows you how to determine training requirements, and teach to the level of your class. It discusses the types, principles, and procedures of teaching and learning, and covers the use of effective training aids and devices. The purpose and principles of testing as well as test construction are covered. Included are chapters on safety, legal considerations, and computers. 5th Edition (1990), 326 pages, addresses NFPA 1041, levels I & II.

LEADERSHIP IN THE FIRE SERVICE

was created from the series of lectures given by Robert F. Hamm to assist in leadership development. It provides the foundation for getting along with others, explains how to gain the confidence of your personnel, and covers what is expected of an officer. Included is information on supervision, evaluations, delegating, and teaching. Some of the topics include: the successful leader today, a look into the past may reveal the future, and self-analysis for officers. 1st Edition (1967), 132 pages.

FIRE SERVICE ORIENTATION AND TERMINOLOGY

Fire Service Orientation and Indoctrination has been revised. It has a new name and a new look. Keeping the best of the old — traditions, history, and organization — this new manual provides a complete dictionary of fire service terms. To be used in conjunction with **Essentials of Fire Fighting** and the other IFSTA manuals. 3rd Edition (1993), addresses NFPA 1001.

PRIVATE FIRE PROTECTION AND DETECTION

provides a means by which fires may be prevented or attacked in their incipient phase and/or controlled until the fire brigade or public fire protection can arrive. This second edition covers information on automatic sprinkler systems, hose standpipe systems, fixed fire pump installations, portable fire extinguishers, fixed special agent extinguishing systems, and fire alarm and detection systems. Information on the design, operation, maintenance, and inspection of these systems and equipment is provided. 2nd Edition (1994).

PUBLIC FIRE EDUCATION

provides valuable information for ending public apathy and ignorance about fire. This manual gives you the knowledge to plan and implement fire prevention campaigns. It shows you how to tailor the individual programs to your audience as well as the time of year or specific problems. It includes working with the media, resource exchange, and smoke detectors. 1st Edition (1979), 169 pages, helpful for NFPA 1021 and 1031.

FIRE DEPARTMENT PUMPING APPARATUS

is the Driver/Operator's encyclopedia on operating fire pumps and pumping apparatus. It covers pumpers, tankers (tenders), brush apparatus, and aerials with pumps. This comprehensive volume explains safe driving techniques, getting maximum efficiency from the pump, and basic water supply. It includes specification writing, apparatus testing, and extensive appendices of pump manufacturers. 7th Edition (1989), 374 pages, addresses NFPA 1002.

STUDY GUIDE FOR PUMPING APPARATUS

The companion study guide in question and answer format. 1990, 100 pages.

FIRE SERVICE RESCUE PRACTICES

is a comprehensive training text for firefighters and fire brigade members that expands proficiency in moving and removing victims from hazardous situations. This extensively illustrated manual includes rescuer safety, effects of rescue work on victims, rescue from hazardous atmospheres, trenching, and outdoor searches. 5th Edition (1981), 262 pages, addresses NFPA 1001.

RESIDENTIAL SPRINKLERS A PRIMER

outlines United States residential fire experience, system components, engineering requirements, and issues concerning automatic and fixed residential sprinkler systems. Written by Gary Courtney and Scott Kerwood and reprinted from *Speaking of Fire*. An excellent reference source for any fire service library and an excellent supplement to **Private Fire Protection.** 1st Edition (1986), 16 pages.

FIRE DEPARTMENT OCCUPATIONAL SAFETY

addresses the basic responsibilities and qualifications for a safety officer and the minimum requirements and procedures for a safety and health program. Included in this manual is an overview of establishing and implementing a safety program, physical fitness and health considerations, safety in training, fire station safety, tool and equipment safety and maintenance, personal protective equipment, en- route hazards and response, emergency scene safety, and special hazards. 2nd Edition (1991), 366 pages, addresses NFPA 1500, 1501.

SALVAGE AND OVERHAUL

covers planning salvage operations, equipment selection and care, as well as describing methods and techniques for using salvage equipment to minimize fire damage caused by water, smoke, heat, and debris. The overhaul section includes methods for finding hidden fire, protection of fire cause evidence, safety during overhaul operations, and restoration of property and fire protection systems after a fire. 7th Edition (1985), 225 pages, addresses NFPA 1001.

SELF-CONTAINED BREATHING APPARATUS

contains all the basics of SCBA use, care, testing, and operation. Special attention is given to safety and training. The chapter on Emergency Conditions Breathing has been completely revised to incorporate safer emergency methods that can be used with newer models of SCBA. Also included are appendices describing regulatory agencies and donning and doffing procedures for nine types of SCBA. The manual has been thoroughly updated to cover NFPA, OSHA, ANSI, and NIOSH regulations and standards as they pertain to SCBA. 2nd Edition (1991), 360 pages, addresses NFPA 1001.

THE SOURCEBOOK FOR FIRE COMPANY TRAINING EVOLUTIONS

provides volunteer and career training officers and company officers with ideas for presenting weekly or monthly training sessions. The book contains plans for more than 50 different training sessions. Each session contains information on the standards that are covered, equipment that is needed, outlines for the presentations and practical exercises, and a listing of pertinent resources and training materials. The sessions cover basic fire fighting, apparatus operation, company evolutions, indoor sessions that can be used on rainy days, and competitive exercises with a practical training value. 1st Edition (1994) 238 pages.

STUDY GUIDE FOR SELF-CONTAINED BREATHING APPARATUS

The companion study guide in question and answer format. 1991, 131 pages.

FIRE STREAM PRACTICES

brings you an all new approach to calculating friction loss. This carefully written text covers the physics of fire and water; the characteristics, requirements, and principles of good streams; and fire fighting foams. **Streams** includes formulas for the application of fire fighting hydraulics, as well as actions and reactions created by applying streams under a variety of circumstances. The friction loss equations and answers are included, and review questions are located at the end of each chapter. 7th Edition (1989), 464 pages, addresses NFPA 1001 and NFPA 1002.

GASOLINE TANK TRUCK EMERGENCIES

provides emergency response personnel with background information, general procedures, and response guidelines to be followed when responding to and operating at incidents involving MC-306/DOT 406 cargo tank trucks. Specific topics include: incident management procedures, site safety considerations, methods of product transfer, and vehicle uprighting considerations. 1st Edition (1992), 51 pages, addresses NFPA 472.

FIRE SERVICE VENTILATION

presents the principles and practices of ventilation. The manual describes and illustrates the safe operations related to ventilation, products of combustion, elements and situations that influence the ventilation process, ventilation methods and procedures, and tools and mechanized equipment used in ventilation. The manual includes chapter reviews, a glossary, and applicable safety considerations. 7th Edition (1994), addresses NFPA 1001.

FIRE SERVICE PRACTICES FOR VOLUNTEER AND SMALL COMMUNITY FIRE DEPARTMENTS

presents those training practices that are most consistent with the activities of smaller fire departments. Consideration is given to the limitations of small community fire department resources. Techniques for performing basic skills are explained, accompanied by detailed illustrations and photographs. 6th Edition (1984), 311 pages.

WATER SUPPLIES FOR FIRE PROTECTION

acquaints you with the principles, requirements, and standards used to provide water for fire fighting. Rural water supplies as well as fixed systems are discussed. Abundant photographs, illustrations, tables, and diagrams make this the most complete text available. It includes requirements for size and carrying capacity of mains, hydrant specifications, maintenance procedures conducted by the fire department, and relevant maps and record-keeping procedures. Review questions at the end of each chapter. 4th Edition (1988), 268 pages, addresses NFPA 1001, NFPA 1002, and NFPA 1031, levels I & II.

CURRICULUM PACKAGES

COMPANY OFFICER

A competency-based teaching package with 17 lessons as well as classroom and practical activities to teach the student the information and skills needed to qualify for the position of Company Officer. Corresponds to **Fire Department Company Officer**, 2nd Edition.

The Package includes the Company Officer Instructor's Guide (the how, what, and when to teach); the Student Guide (a workbook for group instruction); and 143 full-color overhead transparencies.

ESSENTIALS CURRICULUM PACKAGE

A competency-based teaching package with 19 chapters and 22 lessons as well as classroom and practical activities to teach the student the information and skills needed to qualify for the position of Fire Fighter I or II. Corresponds to **Essentials of Fire Fighting**, 3rd Edition.

The Package includes the Essentials Instructor's Guide (the how, what, and when to teach); the Student Guide (a workbook for group instruction); and 445 full-color overhead transparencies.

LEADERSHIP

A complete teaching package that assist the instructor in teaching leadership and motivational skills at the Company Officer level. Each lesson gives an outline of the subject matter to be covered, approximate time required to teach the material, specific learning objectives, and references for the instructor's preparation. Sources for suggested films and videotapes are included.

TRANSLATIONS

LO ESENCIAL EN EL COMBATE DE INCENDIOS

is a direct translation of **Essentials of Fire Fighting**, 2nd edition. Please contact your distributor or FPP for shipping charges to addresses outside U.S. and Canada. 444 pages.

PRACTICAS Y TEORIA PARA BOMBEROS

is a direct translation of **Fire Service Practices for Volunteer and Small Community Fire Departments**, 6th edition. Please contact your distributor or FPP for shipping charges to addresses outside U.S. and Canada. 347 pages.

OTHER ITEMS

TRAINING AIDS

Fire Protection Publications carries a complete line of videos, overhead transparencies, and slides. Call for a current catalog.

NEWSLETTER

The nationally acclaimed and award-winning newsletter, *Speaking of Fire*, is published quarterly and available to you free. Call today for your free subscription.

COMMENT SHEET

DATE _____ NAME _____

ADDRESS _____

ORGANIZATION REPRESENTED _____

CHAPTER TITLE _____ NUMBER _____

SECTION/PARAGRAPH/FIGURE _____ PAGE _____

1. Proposal (include proposed wording or identification of wording to be deleted),
     OR PROPOSED FIGURE:

2. Statement of Problem and Substantiation for Proposal:

RETURN TO: IFSTA Editor   SIGNATURE _____
              Fire Protection Publications
              Oklahoma State University
              Stillwater, OK 74078

Use this sheet to make any suggestions, recommendations, or comments. We need your input to make the manuals as up to date as possible. Your help is appreciated. Use additional pages if necessary.